人工智能与大数据系列

深度学习

从Python到TensorFlow应用实战

叶虎◎编著

清华大学出版社
北京

内容简介

本书全面介绍深度学习中的卷积神经网络结构、学习原理、代码实现、API 调用等基本知识，重点介绍开发深度学习应用所需要的 Python 技术基础以及 TensorFlow 深度学习库，并以文本分类和语音识别为例说明 TensorFlow 的应用场景。

本书可供对 TensorFlow 比较熟悉并且对机器学习有所了解的开发人员、科技工作者和研究人员参考，也可作为高等院校计算机、软件工程等专业高年级本科生与研究生的教材。

本书封面贴有清华大学出版社防伪标签，无标签者不得销售。
版权所有，侵权必究。侵权举报电话：010-62782989　13701121933

图书在版编目（CIP）数据

深度学习：从 Python 到 TensorFlow 应用实战/叶虎编著. —北京：清华大学出版社，2020.1
（人工智能与大数据系列）
ISBN 978-7-302-54565-1

Ⅰ．①深… Ⅱ．①叶… Ⅲ．①机器学习 Ⅳ．①TP181

中国版本图书馆 CIP 数据核字（2019）第 290387 号

责任编辑：张　敏
封面设计：杨玉兰
责任校对：徐俊伟
责任印制：丛怀宇

出版发行：清华大学出版社
网　　址：http://www.tup.com.cn, http://www.wqbook.com
地　　址：北京清华大学学研大厦 A 座　　邮　编：100084
社 总 机：010-62770175　　邮　购：010-62786544
投稿与读者服务：010-62776969, c-service@tup.tsinghua.edu.cn
质量反馈：010-62772015, zhiliang@tup.tsinghua.edu.cn

印 装 者：三河市金元印装有限公司
经　　销：全国新华书店
开　　本：186mm×240mm　　印　张：11.5　　字　数：251 千字
版　　次：2020 年 4 月第 1 版　　印　次：2020 年 4 月第 1 次印刷
定　　价：69.80 元

产品编号：083734-01

前言 Preface

随着人工智能在围棋、德州扑克、人脸识别等方面战胜人类，机器学习等人工智能技术显示了越来越重要的价值。

深度学习（深度结构化学习）是基于一组算法的机器学习的一个分支，这些算法试图通过使用具有复杂结构或由多个非线性变换组成的处理层来对数据中的高层次抽象进行建模。

机器学习技术随着数据集的增长、算法及其实现的改进、硬件性能的提升而持续发展。TensorFlow 作为一种主流的机器学习库让机器学习开发变得越来越容易。

随着可获得文本和语音数据的日益增多，自然语言处理技术在生产和生活中发挥越来越重要的作用。本书介绍如何使用流行的 TensorFlow 进行自然语言处理，并介绍流行的 Python 语言以及使用 Python 开发 TensorFlow 应用。

本书共分 5 章：第 1 章介绍开发深度学习应用所需要的 Linux 基础以及在 Linux 和 Windows 操作系统下搭建开发环境；第 2 章介绍 Python 编程语言基础；第 3 章介绍搭建深度学习框架开发环境，使用 TensorFlow 实现语音识别及 TensorFlow 中的联邦学习；第 4 章介绍通过 TensorFlow.NET 和 TensorFlowSharp 在 C#中使用 TensorFlow 的方法；第 5 章介绍如何使用网格计算引擎 Slurm 构建 Linux 高性能计算集群和如何实现 TensorFlow 在 Slurm 集群的运行。

书中的部分示例采用 Java 或 C#编程语言编写，不熟悉 Java 或者 C#语言的读者可以参考猎兔搜索团队编写的 Java 或者 C#相关入门书籍。

本书适合需要具体实现机器学习应用的开发人员或者对人工智能等相关领域感兴趣的人员参考。

感谢早期合著者、合作伙伴、员工、学员、读者的支持。技术的融合与创新无止境，欢迎一起探索！

在本书的编写过程中，笔者虽尽可能地将清晰的论述呈现给读者，但也难免有疏漏和不妥之处，敬请读者不吝指正。

作者
2020 年 1 月

目 录
Contents

第1章 深度学习快速入门 ··· 1
　1.1　各种深度学习应用 ··· 1
　1.2　准备开发环境 ··· 2
　　　1.2.1　Linux基础 ·· 2
　　　1.2.2　Micro编辑器 ··· 5
　　　1.2.3　Shell基础 ·· 5
　　　1.2.4　Linux下安装Python ·· 8
　　　1.2.5　选择Python版本 ·· 9
　　　1.2.6　使用AWK ·· 9
　　　1.2.7　Windows下安装Python ·································· 10
　　　1.2.8　搭建PyDev集成开发环境 ································ 11
　1.3　体验TensorFlow文本分类 ·· 12
　　　1.3.1　安装TensorFlow ··· 12
　　　1.3.2　实现文本分类 ·· 14
　1.4　本章小结 ··· 16

第2章 Python编程语言基础 ··· 17
　2.1　变量 ··· 17
　2.2　注释 ··· 17
　2.3　简单数据类型 ··· 18
　　　2.3.1　数值 ··· 18
　　　2.3.2　字符串 ·· 20
　　　2.3.3　数组 ··· 22
　2.4　字面值 ··· 22
　2.5　控制流 ··· 23
　　　2.5.1　if语句 ··· 23

2.5.2 循环 ······ 24
2.6 列表 ······ 25
2.7 元组 ······ 28
2.8 集合 ······ 30
2.9 字典 ······ 30
2.10 位数组 ······ 31
2.11 模块 ······ 32
2.12 函数 ······ 33
2.13 print 函数 ······ 35
2.14 正则表达式 ······ 37
2.15 文件操作 ······ 39
 2.15.1 读写文件 ······ 40
 2.15.2 重命名文件 ······ 41
 2.15.3 遍历文件 ······ 41
2.16 使用 pickle 模块序列化对象 ······ 42
2.17 面向对象编程 ······ 42
2.18 命令行参数 ······ 44
2.19 数据库 ······ 45
2.20 JSON 格式 ······ 46
2.21 日志记录 ······ 46
2.22 异常处理 ······ 48
2.23 通过 PyJNIus 使用 Java ······ 48
2.24 本章小结 ······ 49

第 3 章 语音识别中的深度学习 ······ 50
3.1 神经网络基础 ······ 50
 3.1.1 实现深度前馈网络 ······ 52
 3.1.2 计算过程 ······ 61
3.2 卷积神经网络 ······ 67
3.3 语音识别语料库 ······ 73
 3.3.1 TIMIT语料库 ······ 73
 3.3.2 LibriSpeech语料库 ······ 74

3.3.3 中文语料库 …… 74
3.4 搭建深度学习框架开发环境 …… 75
  3.4.1 安装Clang …… 75
  3.4.2 构建配置 …… 79
  3.4.3 configure脚本 …… 80
  3.4.4 静态代码分析 …… 82
  3.4.5 LLDB调试 …… 83
  3.4.6 使用Cygwin模拟环境 …… 86
  3.4.7 使用CMake构建项目 …… 86
  3.4.8 使用Gradle构建项目 …… 87
  3.4.9 Jenkins实现持续集成 …… 92
3.5 TensorFlow 识别语音 …… 92
  3.5.1 使用Keras …… 92
  3.5.2 安装TensorFlow …… 94
  3.5.3 安装TensorFlow的Docker容器 …… 96
  3.5.4 使用TensorFlow …… 97
  3.5.5 一维卷积 …… 137
  3.5.6 二维卷积 …… 139
  3.5.7 膨胀卷积 …… 141
  3.5.8 TensorFlow实现简单的语音识别 …… 142
  3.5.9 NumPy提取语音识别特征 …… 144
  3.5.10 Numba …… 147
3.6 端到端深度学习 …… 148
3.7 Dropout 解决过度拟合问题 …… 148
3.8 NumPy 中的矩阵运算 …… 151
3.9 说话者识别 …… 152
3.10 联邦学习 …… 154
3.11 本章小结 …… 160

第4章 C#开发深度学习应用 161
4.1 使用 TensorFlow.NET …… 161

4.2　使用 TensorFlowSharp ··· 163
　　4.3　本章小结 ·· 164
第 5 章　Slurm 并行训练 ·· 165
　　5.1　网格计算引擎 Slurm 简介 ··· 165
　　　　5.1.1　安装Slurm ·· 166
　　　　5.1.2　Slurm脚本编程 ··· 171
　　5.2　TensorFlow 集群 ·· 172
　　5.3　本章小结 ·· 173
参考文献 ··· 174

第 1 章 深度学习快速入门

源于人工神经网络的深度学习技术随着可用于机器学习数据的积累而迅速发展。深度学习方法可以从大量的训练数据中自动学习出实例的特征，在语音识别、图像识别、自然语言处理等领域得到了广泛的应用。目前已有一些实现深度学习的框架，其中 TensorFlow 是一个流行的框架。

1.1 各种深度学习应用

目前，深度学习技术已应用于安防监控领域识别火灾警情和机器翻译多国语言文字等。

电话机器人首先做好几个相对固定的回答录音，然后识别客户回答的几个关键词，返回相应的应答路由。

JSON 格式的例子如下：

```
"keys": [
    "你是谁",
    "哪位",
    "你叫什么",
    "姓什么",
    "怎么称呼",
    "贵姓"
],
"response": [
    [
    "我姓王，你叫我小王就可以了",
    "是这样的，免贵姓王，您可以叫我小王"
    ]
```

可以采用模板引擎来编写应答。已有各种编程语言实现的模板引擎，例如用 Java 语言实现的 FreeMarker（https://freemarker.apache.org/）或者用 Python 语言实现的 Cheetah3（http://cheetahtemplate.org/）。

另外，医疗系统可以根据语音识别患者主诉的结果，实现病历的自动填写。

1.2 准备开发环境

当前，很多语音识别应用都是在 Linux 操作系统下开发的。Linux 是围绕 Linux 内核构建的免费和开源软件操作系统系列。Linux 来源于 UNIX，是 UNIX 操作系统的开放源代码实现。Linux 发行版是一个基于 Linux 内核的由软件集合构成的操作系统。通常，Linux 用户通过下载一个 Linux 发行版获得操作系统。Linux 有一些常用的发行版，如 CentOS 和 Ubuntu 等版本。Ubuntu 是由 Canonical 公司开发的基于 Debian 的开源 Linux 操作系统；CentOS 是 Red Hat Enterprise Linux 的免费复制版。本节首先介绍在 Ubuntu 和 CentOS 下安装 Python，然后介绍在 Linux 下开发 Python 应用的编辑器 Micro。

在 Windows 下，可以使用 Sublime 这样的文本编辑器编写 Python 代码，也可以使用 PyDev 或者 PyCharm 集成开发环境。

1.2.1 Linux 基础

有些语音识别系统运行在 Linux 服务器中。为了远程登录 Linux 服务器，可以安装 KiTTY（https://www.fosshub.com/KiTTY.html）。在 KiTTY 的配置界面输入 IP 地址，用户名和密码后登录 Linux 服务器。如果是用 root 账户登录，则终端提示符是#，否则终端提示符是$。

查看 Ubuntu 操作系统版本号：

```
$cat /etc/issue
Ubuntu 18.04 LTS \n \l
```

或者：

```
$ lsb_release -r
Release:        18.04
```

获取 Ubuntu 的代号：

```
$ lsb_release -c
    Codename:       bionic
```

ls 命令用于列出当前目录下的文件。有的命令比较长，为了快速输入，可以用 Tab 键补全命令。History 是显示历史命令，用上箭头选择最近运行过的命令可再次执行。

可以使用支持 SSH 协议的终端仿真程序 SecureCRT 连接到远程 Linux 服务器。因为可以保存登录密码，所以比较方便。除了 SecureCRT，还可以使用开源软件 PuTTY（http://www.chiark.greenend.org.uk/~sgtatham/putty），以及可以保存登录密码的 PuTTY Connection Manager。

在终端启动的进程断开连接后进程会停止运行。为了让进程继续运行，可以使用 nohup 命令。

如果需要安装软件，可以下载对应的 RPM 安装包，然后使用 RPM 安装。但操作系统对应的 RPM 安装包找起来往往比较麻烦。一个软件包可能依赖其他的软件包。为了安装一个软件可能需要下载其他几个它所依赖的软件包。

为了简化安装操作，可以使用黄狗升级管理器，一般简称 yum。yum 会自动计算出程序之间的相互关联性，并且计算出完成软件包的安装需要哪些步骤。这样在安装软件时，不会再被那些关联性问题所困扰。

yum 软件包管理器自动从网络下载并安装软件。yum 有点类似 360 软件管家，但是不会有商业倾向的推销软件。例如安装支持 wget 和 rzsz 命令的软件：

```
#yum install wget
#yum install lrzsz
```

Windows 格式文本文件的换行符为\r\n，而 Linux 文件的换行符为\n。

dos2unix 是将 Windows 格式文件转换为 Linux 格式的实用命令。dos2unix 命令其实就是将文件中的\r\n 转换为\n。

开发语音识别系统的过程中，可能会用到大量的数据文件。如需要在 Linux 操作系统上维护同一文件的两份或多份副本，除了保存多份单独的物理文件副本之外，还可以采用保存一份物理文件副本和多个虚拟副本的方法。这种虚拟的副本就称为链接。链接是目录中指向文件真实位置的占位符。在 Linux 中有两种不同类型的文件链接：符号链接和硬链接。

符号链接就是一个实实在在的文件，它指向存放在虚拟目录结构中某个地方的另一个文件。这两个通过符号链接在一起的文件，彼此的内容并不相同。

如果现有空间不够用，可以增加存储设备后扩容。首先用 lsblk 命令查看现有空间情况。在一个 Linux 账号中显示如下：

```
[root@localhost ~]#lsblk
NAME     MAJ:MIN RM  SIZE RO TYPE MOUNTPOINT
sr0       11:0    1 1024M  0 rom
xvda     202:0    0  100G  0 disk
├─xvda1  202:1    0    8G  0 part [SWAP]
```

```
└─xvda2   202:2   0   32G    0 part /mnt
xvde      202:64  0 1000G    0 disk
```

创建一个要扩展的目录：

```
#mkdir /ext
```

加载文件系统到这个目录下：

```
#mount /dev/xvde /ext
```

确认加载成功：

```
[root@localhost ~]#df -m
Filesystem       1M-blocks    Used    Available   Use%    Mounted on
/dev/xvda2       32752        32496   256         100%    /
devtmpfs         32108        0       32108       0%      /dev
tmpfs            32020        1       32020       1%      /dev/shm
tmpfs            32020        186     31835       1%      /run
tmpfs            32020        0       32020       0%      /sys/fs/cgroup
tmpfs            6374         1       6374        1%      /run/user/0
/dev/xvde        1007801      77      956509      1%      /ext
```

对于大的文件可以使用 wget 命令在后台下载。

```
#wget -bc <path>
```

这里的参数 b 表示在后台运行；参数 c 表示支持断点续传。

可以在 Windows 下编辑文本文件，然后使用 Perl 把 Windows 下的文本文件转换成 Linux 可识别的格式：

```
#perl -p -e 's/\r$//' < winfile.txt > unixfile.txt
```

在 Ubuntu 操作系统下安装 Python3：

```
#apt install python3
```

Ubuntu 系统上默认的 root 密码是随机生成的，而 root 权限是通过 sudo 命令授予的。可以在终端输入命令 sudo passwd，然后输入当前用户的密码，按回车键，终端会提示输入新的密码并确认，此时的密码就是 root 新密码。修改成功后，输入命令 su root，再输入新的密码即可。

可以使用 sed(Stream EDitor)命令查找和替换文件中的文本。例如：

```
#sed -i 's/original/new/g' file.txt
```

命令行参数说明如下：

-i：-i = in-place（保存回原始文件）。

命令字符串说明如下：

s：s = 替换命令。

original：original =描述要替换的单词的正则表达式（或者只是单词本身）。

new：new =用来替换的目标文本。

g：g = global（即替换所有而不仅仅是第一次出现的）。

file.txt：file.txt =文件名。

例如把 cmd.sh 中的 queue.pl 替换成为 run.pl，替换结果输出到 cmd.local.sh 文件。

```
#sed 's/queue.pl/run.pl/g' cmd.sh > cmd.local.sh
```

1.2.2 Micro 编辑器

为了方便在服务器端开发 Python、Perl、Shell、C++相关应用，可以采用 Micro（https://github.com/zyedidia/micro）这样的终端文本编辑器。

可以在/home/soft/micro 目录下运行：

```
#curl https://getmic.ro | bash
```

设置成在任意路径均可运行 Micro：

```
#cd /usr/bin
#sudo ln -s /home/soft/micro/micro micro
```

或者编辑/etc/profile 文件，增加 micro 所在的路径到 PATH 环境变量/home/soft/micro。

```
#./micro /etc/profile
```

增加如下行：

```
export PATH=/home/soft/micro:$PATH
```

可以使用它编辑配置文件：

```
#./micro run.pl
```

输入：

```
die "run.pl: Hello Error";
```

这里的 die 表示终止脚本运行，并显示出 die 后面的双引号中的内容。

保存文件后，按 Ctrl+Q 组合键退出。

1.2.3 Shell 基础

Shell 是用户和 Linux 内核之间的接口程序。在命令行提示符下输入的每个命令都由 Shell 先解释然后传给 Linux 内核。

Shell 是一个命令语言解释器，拥有自己内建的 Shell 命令集。此外，Shell 也能被系统中其他有效的 Linux 实用程序和应用程序所调用。

Shell 具有如下主要功能。

命令解释功能：将用户可读的命令转换成计算机可理解的命令，并控制命令执行。

输入/输出重定向：操作系统将键盘作为标准输入，将显示器作为标准输出。当这些定向不能满足用户需求时，用户可以在命令中用符号">"或"<"重新定向。

管道处理：利用管道将一个命令的输出送入另一个命令，实现多个命令组合完成复杂命令的功能。

系统环境设置：用 Shell 命令设置环境变量，维护用户的工作环境。

程序设计语言：Shell 命令本身可以作为程序设计语言，将多个 Shell 命令组合起来，编写能实现系统或用户所需功能的程序。

Shell 有很多种，如 zshell 和 fish。这里介绍 Bash（Bourne Again Shell）。

可以使用 ps 命令查看当前使用的是哪种 shell。

```
#ps
    PID TTY          TIME CMD
19977 pts/1    00:00:00 bash
50063 pts/1    00:00:00 ps
```

在屏幕上打印"Hello"：

```
echo "Hello"
```

将 ABC 分配给 a：

```
a=ABC
```

输出 a 的值：

```
echo $a
```

在屏幕上打印 ABC。

将 ABC.log 分配给 b：

```
b=$a.log
```

输出 b 的值：

```
#echo $b
ABC.log
```

把文件"ABC.log"的内容写入到 testfile：

```
cat $b > testfile
```

"指令 --help"会输出帮助信息。

可以把重复执行的 Shell 脚本写入一个文本文件。在 Linux 中，文件扩展名不作为系统识别文件类型的依据，但是可以作为我们识别文件的依据，可以简单地将脚本文件名以.sh 结尾。

在 Linux 下，可以通过 vi 命令创建一个诸如 script.sh 的文件：vi script.sh。创建好脚本文件后就可以在文件内按脚本语言要求的格式编写脚本程序了。

在创建的脚本文件中输入以下代码并保存退出：

```
#! /bin/bash
echo "hello world!"
```

添加脚本文件的可执行运行权限 chmod 777 script.sh，之后运行文件./script.sh 得到结果：

```
hello world!
```

Shell 脚本中用#表示注释，相当于 C 语言的//注释。但如果#位于第一行开头，并且是#!（称为 Shebang）则例外，它表示该脚本使用后面指定的解释器/bin/sh 解释执行。每个脚本程序必须在开头包含这个语句。

使用参数 n 检查语法错误，例如：

```
#bash -n ./test.sh
```

如果 Shell 脚本有语法错误，则会提示错误所在行；否则，不输出任何信息。

if 语句的语法是：

```
if [ condition ] then
    command1
elif               #和 else if 等价
    then
        command2
    else
        default-command
fi
```

这里的 fi 就是 if 反过来写。

例如，为了判断某个命令是否存在，可以使用如下的格式：

```
if which programname >/dev/null; then
    echo exists
else
    echo does not exist
fi
```

判断 yum 是否存在的例子：

```
if which yum >/dev/null; then
    echo "exists"
else
```

```
    echo "does not exist"
fi
```

case 语句的语法是：

```
case 字符串 in
    模式 1)
        语句
        ;;
    模式 2)
        语句
        ;;
    *)
        默认执行的语句
        ;;
esac
```

这里的 esac 就是 case 反过来写。例如：

```
extension="png"
case "$extension" in
    "jpg"|"jpeg")
        echo "It's image with jpeg extension."
        ;;
    "png")
        echo "It's image with png extension."
        ;;
    "gif")
        echo "Oh, it's a giphy!"
        ;;
    *)
        echo "Woops! It's not image!"
        ;;
esac
```

这里使用"|"把"jpg"和"jpeg"这两个模式连接到了一起。

1.2.4　Linux 下安装 Python

首先检查 Python 3 是否已经正确安装，以及所使用的版本号：

```
#python3 -V
Python 3.4.5
```

检查 Python 3 所在的路径：

```
#which python3
```

```
/usr/bin/python3
```

如果使用 CentOS，可以使用 yum 安装 Python 3。首先查找可供安装的 Python 版本：

```
#yum search python3
```

然后安装想要的版本：

```
#yum install python36
```

如果使用 Ubuntu 操作系统，首先运行以下命令更新软件包列表并将所有系统软件升级到可用的最新版本：

```
#sudo apt-get update && sudo apt-get -y upgrade
```

然后安装 pip 包管理系统：

```
#sudo apt-get install python3-pip
```

1.2.5 选择 Python 版本

Linux 系统中有可能同时存在多个可用的 Python 版本，每个 Python 版本都对应一个可执行二进制文件。可以使用 ls 命令来查看系统中有哪些 Python 的二进制文件可供使用。

```
$ ls /usr/bin/python*
```

python 命令执行 Python 2。可以使用 python3 命令执行 Python 3。如何使用 python 命令执行 Python 3？

一种简单安全的方法是使用别名。将如下命令放入~/.bashrc 或~/.bash_aliases 文件中：

```
alias python=python3
```

最好在终端中使用'python3'命令，在 Python 3.x 文件中使用 shebang 行'#!/usr/bin/env python3'。

1.2.6 使用 AWK

典型的 AWK 程序充当过滤器。它从标准输入读取数据，并输出标准输出的过滤数据。它一次读取数据的一个记录。默认情况下，一次读取一行文本。每次读取记录时，AWK 自动将记录分隔到字段中。字段在默认情况下也是由空格分隔的。每个字段被分配给一个变量，该变量有一个数字名称。变量$1 是第一个字段，$2 是第二个字段，以此类推。$0 表示整个记录。此外，还设置了一个名为 NF 的变量，其中包含在记录中检测到的字段的数目。

来试试一个很简单的例子。过滤 ls 命令的输出：

```
#ls -l ./ | awk '{print $0}'
```

显示文本文件 nohup.out 匹配（含有）字符串"sun"的所有行：

```
#awk '/sun/{print}' nohup.out
```

由于显示整个记录（全行）是 awk 的默认动作，因此可以省略 action 项。

例如，得到 Python 的版本号：

```
#python 2>&1 --version | awk '{print $2}'
2.7.5
```

这里的 2>&1 意思是：把标准错误重定向到标准输出。

1.2.7　Windows 下安装 Python

在图形化用户界面出现之前，人们就是用命令行来操作计算机的。Windows 命令行是通过 Windows 系统目录下的 cmd.exe 执行的。执行这个程序最直接的方式是找到这个程序，然后双击。但 cmd.exe 并没有一个桌面的快捷方式，所以这样太麻烦。

可以在开始菜单的运行窗口直接输入程序名，回车后运行这个程序。打开"开始"→"运行"，这样就会打开资源管理器中的运行程序窗口。或者使用快捷键——窗口键+R，打开运行程序窗口。总之，输入要运行的程序名 cmd 后单击"确定"按钮，出现命令提示窗口。因为能够通过这个黑屏的窗口直接输入命令来控制计算机，所以也称为控制台窗口。

开始的路径往往是 C:\Users\Administrator。就像公园的地图上往往会标出游客的当前位置。Windows 命令行也有个当前目录的概念。这个 C:\Users\Administrator 就是当前路径。

可以用 cd 命令改变当前路径，例如改变到 C:\Python\Python27 路径。

```
C:\Users\Administrator>cd C:\Python\Python27
```

系统约定从指定的路径找可执行文件。这个路径通过 PATH 环境变量指定。环境变量是一个"变量名=变量值"的对应关系，每一个变量都有一个或者多个值与之对应。如果是多个值，则这些值之间用分号隔开。例如 PATH 环境变量可能对应这样的值："C:\Windows\system32;C:\Windows"。表示 Windows 会从 C:\Windows\system32 和 C:\Windows 两个路径找可执行文件。

设置或者修改环境变量的具体操作步骤是：首先在 Windows 桌面右键"计算机"，在弹出的"属性"对话框中选择"高级"→"环境变量"，然后设置用户变量，或者系统变量，接着再设置环境变量 PATH 的值。

需要重新启动命令行才能让环境变量设置生效。为了检查环境变量是否设置正确，可以在命令行中显示指定环境变量的值。需要用到 echo 命令。echo 命令用来显示一段文字。

```
C:\Users\Administrator>echo Hello
```

执行上面的命令将在命令行输出：Hello。

如果要引用环境变量的值，可以用前后两个百分号把变量名包围起来："%变量名%"。echo 命令用来显示一个环境变量中的值。

```
C:\Users\Administrator>echo %PATH%
```

假设把 Python 安装在 D:\Python\Python27 目录下，则可以在计算机属性中手工设置 PATH 环境变量，然后检查环境变量的值：

```
C:\Users\Administrator.PC-201509301458>echo %PATH%
C:\Windows\system32;C:\Windows;C:\Windows\System32\Wbem;C:\Windows\System32\WindowsPowerShell\v1.0\;D:\apache-maven-3.5.2\bin;D:\Python\Python27
```

用如下命令检查 Python 是否正确安装，以及所使用的版本号：

```
>python -version
```

或者用 where 命令检查系统是否已经安装了 Python：

```
C:\Users\Administrator>where python
C:\Python27\python.exe
D:\cygwin64\bin\python
D:\Programs\Python\Python36\python.exe
```

如果没有安装 Python，则可以使用 Chocolatey（https://chocolatey.org）安装 Python 3：

```
>choco install python3
```

1.2.8　搭建 PyDev 集成开发环境

在 Windows 下可以使用 PyDev（https://github.com/fabioz/Pydev）开发 Python 应用。

首先检查 Windows 操作系统中是否已经安装了 JDK：

```
>where javac
C:\Program Files\Java\jdk1.8.0_181\bin\javac.exe
```

如果没有，则可以使用软件包管理器 Chocolatey（https://chocolatey.org/）安装 JDK：

```
>choco install jdk
```

从 Eclipse 官方网站下载 Eclipse。

解压 pydev 插件后，会发现有 features、plugins 两个文件夹，然后把 features 文件夹下的所有文件移到 D:\Program Files\eclipse\features 目录下，把 plugins 文件夹下的所有文件移到 D:\Program Files\eclipse\plugins 目录下。

注意：D:\Program Files\eclipse 为 Eclipse 安装目录。

重启 Eclipse，在 windows->preferences 出现 PyDev 配置项，表示 PyDev 插件安装成功。

创建新项目，项目属性改成 UTF-8 编码。创建 Python 模块：

```
'''
Created on 2019 年 3 月 6 日

@author: Administrator
'''
print('hi')
```

单击工具栏的 Run 按钮运行。控制台视图显示输出结果：

```
hi
```

这里使用了 print 函数输出一个字符串到控制台。

Eclipse 默认是英文界面，如果习惯用中文界面，可以从 http://www.eclipse.org/babel/downloads.php 下载支持中文的语言包。

如果想要切换回英文开发环境，则可以使用命令行进入 Eclipse 主目录后输入：

```
eclipse -nl en
```

切换回简体中文：

```
eclipse -nl zh_CN
```

然后安装插件，配置路径。

插件通过 Eclipse 的软件安装机制安装。从"帮助"菜单中选择"安装新软件"。输入网址 http://pydev.org/updates，并选择 PyDev 执行 next 开始安装，安装完需要重启 Eclipse。

必须配置 PyDev 才能与 Eclipse 和 Python 设置一起正常工作。从 Eclipse 主菜单中选择 Window→Preferences。这将打开"首选项"对话框。选择 Pydev→Interpreter - Python：

"Quick Auto-Config" 按钮可以按照用户希望的方式设置 Eclipse，但前提是选择了某个版本的 Python 3。如果找到早期版本的 Python，请确保使用"删除"按钮将其删除，然后手动选择 python.exe 文件所在的路径。

1.3 体验 TensorFlow 文本分类

以文本分类这个经典问题来体验 TensorFlow。

1.3.1 安装 TensorFlow

这里介绍在 Linux 操作系统下安装 TensorFlow。当前推荐使用 Ubuntu 发行版本。

当用户最终让自己的 Linux 操作系统正常运行以后，请打开一个终端并安装一些必要的软件。

git：分布式版本控制系统。

wget：使用 HTTP、HTTPS 和 FTP 协议进行数据传输。

必须安装的软件包括：

awk：编程语言，用于搜索和处理文件和数据流中的模式。

bash：UNIX Shell 和脚本编程语言。

grep：逐行处理文本并打印与指定模式匹配的任何行。

make：从源代码自动构建可执行程序和库。

bazel：从源代码自动构建 TensorFlow 可执行程序和库。

perl：动态编程语言，非常适合文本文件处理。

例如，安装 git 可以用如下的命令。

```
#apt-get install git
```

要仅为在 CPU 上使用而安装当前版本：

```
#pip install tensorflow
```

如果要使用支持 CUDA 的 GPU 卡，则安装 TensorFlow 的 GPU 版本：

```
#pip install tensorflow-gpu
```

在交互式环境测试 TensorFlow：

```
>>> import tensorflow as tf
>>> tf.enable_eager_execution()
>>> tf.add(1, 2).numpy()
3
>>> hello = tf.constant('Hello, TensorFlow!')
>>> hello.numpy()
'Hello, TensorFlow!'
```

使用函数 tf.nn.softmax()测试 Tensorflow。例如，有一个 4 维的向量。-1 是最低值，3 是最高值。这些值都归一化为 0~1 的数值。

```
import tensorflow as tf
x = [0., -1., 2., 3.]
softmax_x = tf.nn.softmax(x)
session = tf.Session()
print('x:',x)
print('softmax_x:',session.run(softmax_x))
```

softmax 是逻辑函数的推广，softmax 将任意实数值的 K 维向量"压缩"到[0,1]区间的实数值的 K 维向量，向量中的元素值加起来为 1。

如果 TensorFlow 依赖的 numpy 出错，则可以考虑先卸载 numpy，然后重新安装 Tensorflow。

```
>pip uninstall numpy
>python -m pip install tensorflow
```

可以使用交互式会话测试 Tensorflow：

```
>>> import tensorflow as tf
>>> x = tf.placeholder("float", [2])
>>> a = tf.placeholder(tf.float32, name='a')
>>> b = tf.placeholder(tf.float32, name='b')
>>> c = tf.add(a, b, name='c')
>>> sess = tf.InteractiveSession()
>>> sess.run(c, feed_dict={a: 2.1, b: 1.9})
4.0
```

1.3.2 实现文本分类

在训练文本分类模型之前，必须先准备数据。可以创建一个简单的 JSON 文件来保存训练所需的数据。

```
{
    "time" : ["what time is it?", "how long has it been since we started?", "that's a long time ago", " I spoke to you last week", " I saw you yesterday"],
    "sorry" : ["I'm extremely sorry", "did he apologize to you?", "I shouldn't have been rude"],
    "greeting": ["Hello there!", "Hey man! How are you?", "hi"],
    "farewell": ["It was a pleasure meeting you", "Good Bye.", "see you soon", "I gotta go now."],
    "age": ["what's your age?", "How old are you?", "I'm a couple of years older than her", "You look aged!"]
}
```

以下是示例 JSON 训练的数据文件，其中包含 5 个类别。

```
{
    "time": ["what time is it?", "how long has it been since we started?", "that's a long time ago", " I spoke to you last week", " I saw you yesterday"],
    "sorry": ["I'm extremely sorry", "did he apologize to you?", "I shouldn't have been rude"],
    "greeting": ["Hello there!", "Hey man! How are you?", "hi"],
    "farewell": ["It was a pleasure meeting you", "Good Bye.", "see you soon", "I gotta go now."],
    "age": ["what's your age?", "How old are you?", "I'm a couple of years older than her", "You look aged!"]
}
```

数据加载和预处理：

```python
#用于保存标点符号的表结构
tbl = dict.fromkeys(i for i in range(sys.maxunicode)
                if unicodedata.category(chr(i)).startswith('P'))
#从句子中删除标点符号的方法
def remove_punctuation(text):
    return text.translate(tbl)
#初始化词干分析器
stemmer = LancasterStemmer()
#用于保存从文件中读取的JSON数据的变量
data = None
#读取JSON文件并加载训练数据
with open('data.json') as json_data:
    data = json.load(json_data)
    print(data)
#获取要训练的所有类别的列表
categories = list(data.keys())
words = []
#包含句子中单词和类别名称的元组列表
docs = []
for each_category in data.keys():
    for each_sentence in data[each_category]:
        #从句子中删除标点符号
        each_sentence = remove_punctuation(each_sentence)
        print(each_sentence)
        #从每个句子中提取单词并附加到单词列表中
        w = nltk.word_tokenize(each_sentence)
        print("tokenized words: ", w)
        words.extend(w)
        docs.append((w, each_category))
#词干化并小写化每个单词，然后删除重复项
words = [stemmer.stem(w.lower()) for w in words]
words = sorted(list(set(words)))
print(words)
print(docs)
```

构建一个简单的深度神经网络，并用它来训练模型。

```python
#重置底层图数据
tf.reset_default_graph()
#构建神经网络
net = tflearn.input_data(shape=[None, len(train_x[0])])
net = tflearn.fully_connected(net, 8)
net = tflearn.fully_connected(net, 8)
net = tflearn.fully_connected(net, len(train_y[0]), activation='softmax')
```

```
net = tflearn.regression(net)
#定义模型并设置tensorboard
model = tflearn.DNN(net, tensorboard_dir='tflearn_logs')
#开始训练（应用梯度下降算法）
model.fit(train_x, train_y, n_epoch=1000, batch_size=8, show_metric=True)
model.save('model.tflearn')
```

使用下面的代码测试神经网络文本分类 Python 模型。

```
#用几句话测试一下模型:
#前两个句子在训练集中出现过，最后两个句子没有在训练集中出现过
sent_1 = "what time is it?"
sent_2 = "I gotta go now"
sent_3 = "do you know the time now?"
sent_4 = "you must be a couple of years older then her!"
#一个接收句子的方法，并以可以馈送到tensorflow的形式返回数据
def get_tf_record(sentence):
    global words
    #切分出句子中的词
    sentence_words = nltk.word_tokenize(sentence)
    #返回每个词的词干
    sentence_words = [stemmer.stem(word.lower()) for word in sentence_words]
    #词袋
    bow = [0]*len(words)
    for s in sentence_words:
        for i, w in enumerate(words):
            if w == s:
                bow[i] = 1
    return(np.array(bow))
#开始预测4个句子中每个句子的分类结果
print(categories[np.argmax(model.predict([get_tf_record(sent_1)]))])
print(categories[np.argmax(model.predict([get_tf_record(sent_2)]))])
print(categories[np.argmax(model.predict([get_tf_record(sent_3)]))])
print(categories[np.argmax(model.predict([get_tf_record(sent_4)]))])
```

1.4 本章小结

由 Google 大脑团队开发，供 Google 内部使用的一个深度学习框架于 2015 年命名为 TensorFlow。TensorFlow 可运行于 Linux 或者 Windows、MacOS 等操作系统。

第 2 章 Python 编程语言基础

本章介绍开发深度学习应用所需要的 Python 编程语言基础。

2.1 变量

在 Python 编程语言中，定义变量时不声明类型，但变量在内部是有类型的。变量的类型由函数 type() 得到。例如，在交互式环境中输入代码：

```
>>> a=3
>>> type(a)
<class 'int'>
```

表明变量 a 是整数类型。

2.2 注释

与 Shell 类似，Python 脚本中用#表示注释。但如果#位于第一行开头，并且是#!（称为 Shebang）则例外，它表示该脚本使用后面指定的解释器/usr/bin/python3 解释执行。每个脚本程序只能在开头包含这个语句。

为了能够在源代码中添加中文注释，需要把源代码保存成 UTF-8 格式。例如：

```
#-*- coding: utf-8 -*
```

```
import tensorflow as tf
x = tf.placeholder("float", [2])   #形状是 2
```

2.3 简单数据类型

本节介绍包括数值、字符串和数组在内的简单数据类型。

2.3.1 数值

Python 中有三种不同的数值类型：int（整数）、float（浮点数）和 complex（复数）。

和 Java 或者 C 语言中的 int 类型不同。Python 中的 int 类型是无限精度的。例如：

```
>>> i=3243244444444444444444444444444444444444487976875675676570000000000000000000000000000000000000000000000000000000000000000000000000000564564
>>> i
3243244444444444444444444444444444444444487976875675676570000000000000000000000000000000000000000000000000000000000000000000000000000564564
>>> type(i)
<class 'int'>
```

Python 依据 IEEE 754 标准使用二进制表示 float（浮点数），所以存在表示精度的问题。例如：

```
>>> 0.1 == 0.10000000000000000000001
True
```

可以使用 decimal 模块使用十进制表示完整的小数。例如：

```
>>> import decimal
>>> a = decimal.Decimal('0.1')
>>> b = decimal.Decimal('0.10000000000000000000001')
>>> a == b
False
```

在傅里叶变换中会用到复数。复数在 Python 中是一个基本数据类型（complex）。例如：

```
>>> complex(2,3)
(2+3j)
```

一个复数有一些内置的访问器：

```
>>> z = 2+3j
>>> z.real
2.0
```

```
>>> z.imag
3.0
>>> z.conjugate()
(2-3j)
```

几个内置函数支持复数：

```
>>> abs(3 + 4j)
5.0
>>> pow(3 + 4j, 2)
(-7+24j)
```

标准模块 cmath 具有处理复数的更多功能：

```
>>> import cmath
>>> cmath.sin(2 + 3j)
(9.15449914691143-4.168906959966565j)
```

用于数值运算的算术运算符说明见表 2-1。

表 2-1 算术运算符说明

语 法	数 学 含 义	运算符名字
a+b	$a+b$	加
a-b	$a-b$	减
a*b	$a \times b$	乘法
a/b	$a \div b$	除法
a//b	$\lfloor a \div b \rfloor$	地板除
a%b	$a \bmod b$	模
-a	$-a$	取负数
abs(a)	$|a|$	绝对值
a**b	a^b	指数
math.sqrt(a)	$\sqrt{a}$	平方根

对于"/"运算，就算分子、分母都是 int，返回的也将是浮点数。例如：

```
>>> print(1/3)
0.3333333333333333
```

Python 支持不同的数字类型相加，它使用数字类型强制转换的方式来解决数字类型不一致的问题。也就是说，它会将一个操作数转换成与另一个操作数相同的数据类型。

如果有一个操作数是复数，则另一个操作数被转换为复数：

```
>>> 3.0 + (5+6j)          #非复数转换为复数
```

```
(8+6j)
```

整数转换为浮点数：

```
>>> 6 + 7.0          #非浮点型转浮点型
13.0
```

Python 代码中一般一行就是一条语句，但是可以使用斜杠（\）将一条语句分为多行显示。例子代码如下：

```
>>> a = 1
>>> b = 2
>>> c = 3
>>> total = a + \
b + \
c
>>> total
6
```

2.3.2 字符串

可以使用方法 strip()去掉字符串首尾的空格或者指定的字符。

```
term = "   hi    ";   #去除首尾空格
print(term.strip());
```

使用方法 split()将句子分成单词。例如下面的代码中，Mary 是一个单一的字符串。尽管这是一个句子，但这些词语并没有表示成严谨的单位。为此，需要一种不同的数据类型，即字符串列表，其中每个字符串对应一个单词。使用方法 split()可把句子切分成单词：

```
>>> mary = 'Mary had a little lamb'
>>> mary.split()
['Mary', 'had', 'a', 'little', 'lamb']
```

split()方法根据空格拆分 Mary，返回的结果是 Mary 中的单词列表。此列表包含函数 len()演示的 5 个项目。对于 Mary，函数 len()返回字符串中的字符数（包括空格）。

```
>>> mwords = mary.split()
>>> mwords
['Mary', 'had', 'a', 'little', 'lamb']
>>> len(mwords)          #mwords 中的项目数
5
>>> len(mary)            #字符数
22
```

空白字符包括空格' '、换行符'\ n'和制表符'\ t'等。.split()分隔这些字符的任何组合序列：

```
>>> chom = ' colorless    green \n\tideas\n'
>>> print(chom)
 colorless    green
	ideas

>>> chom.split()
['colorless', 'green', 'ideas']
```

通过提供可选参数，.split('x')可用于在特定子字符串'x'上拆分字符串。如果没有指定'x'，.split()只是在所有空格上分割，如上所示。

```
>>> mary = 'Mary had a little lamb'
>>> mary.split('a')       #根据'a'切分
['M', 'ry h', 'd ', ' little l', 'mb']
>>> hi = 'Hello mother,\nHello father.'
>>> print(hi)
Hello mother,
Hello father.
>>> hi.split()            #没有给出参数：在空格上分割
['Hello', 'mother,', 'Hello', 'father.']
>>> hi.split('\n')        #仅在'\n'上分割
['Hello mother,', 'Hello father.']
```

但是如果想将一个字符串拆分成一个字符列表呢？在 Python 中，字符只是长度为 1 的字符串。函数 list()将字符串转换为单个字母的列表：

```
>>> list('hello world')
['h', 'e', 'l', 'l', 'o', ' ', 'w', 'o', 'r', 'l', 'd']
```

如果有一个单词列表，可以使用.join()方法将它们重新组合成一个单独的字符串。在"分隔符"字符串'x'上调用，'x'.join(y)连接列表 y 中由'x'分隔的每个元素。下面，mwords 中的单词用空格连接回句子字符串：

```
>>> mwords
['Mary', 'had', 'a', 'little', 'lamb']
>>> ' '.join(mwords)
'Mary had a little lamb'
```

也可以在空字符串""上调用该方法作为分隔符。效果是列表中的元素连接在一起，元素之间没有任何内容。下面，将一个字符列表放回到原始字符串中：

```
>>> hi = 'hello world'
>>> hichars = list(hi)
>>> hichars
['h', 'e', 'l', 'l', 'o', ' ', 'w', 'o', 'r', 'l', 'd']
>>> ''.join(hichars)
```

```
'hello world'
```

一个字符串取子串的例子代码如下：

```
>>> x = "Hello World!"
>>> x[2:]
'llo World!'
>>> x[:2]
'He'
>>> x[:-2]
'Hello Worl'
>>> x[-2:]
'd!'
>>> x[2:-2]
'llo Worl'
```

使用函数 ord() 和函数 chr() 实现字符串和整数之间的相互转换：

```
>>> a = 'v'
>>> i = ord(a)
>>> chr(i)
'v'
```

2.3.3 数组

可以使用 array（数组）存储同样数据类型的数值类型。通过 import array 导入 Python 的数组类型，就可以使用 array 类型。

例如：

```
from array import array
node=array('H')          #存储无符号短整型的数组

node.append(12)
```

2.4 字面值

Python 包括如下几种类型的字面值。

数字：整数、浮点数、复数。

字符串：以单引号、双引号或者三引号定义字符串。

布尔值：True 和 False。

空值：None。

有 4 种不同的字面值集合，分别是：列表字面值、元组字面值、字典字面值和集合字面值。示例代码如下：

```
fruits = ["apple", "mango", "orange"]              #列表
numbers = (1, 2, 3)                                #元组
alphabets = {'a':'apple', 'b':'ball', 'c':'cat'}   #字典
vowels = {'a', 'e', 'i' , 'o', 'u'}                #集合

print(fruits)
print(numbers)
print(alphabets)
print(vowels)
```

2.5 控制流

完成一件事情要有流程控制。例如，用洗衣机洗衣服有 3 个步骤：①把脏衣服放进洗衣机；②等洗衣机洗好衣服；③晾衣服。这是顺序控制结构。

顺序执行的代码采用相同的缩进，叫作一个代码块。Python 没有像 Java 或者 C#语言那样采用{}分隔代码块，而是采用代码缩进和冒号来区分代码之间的层次。

缩进的空格数量是可变的，但是所有代码块语句必须包含相同的缩进空格数量。NodePad++这样的文本编辑器支持选择多行代码后，按 Tab 键改变代码块的缩进格式。

控制流用根据运行时的情况来调整语句的执行顺序。流程控制语句可以分为条件语句和迭代语句。

2.5.1 if 语句

当路径不存在时，就创建它，可以使用条件语句来实现。条件语句的一般形式如下：

```
if 条件:
    语句 1
    语句 2...
elif 条件:
    语句 1
    语句 2...
else:
    语句 1
    语句 2...
语句 x
```

例如，判断一个数是否是正数：

```
x = -32.2;
isPositive = (x > 0);
if isPositive:
    print(x, " 是正数");
else:
    print(x, " 不是正数");
```

这里的 if 复合语句，首行以关键字开始，以冒号（:）结束。

使用关系运算符和条件运算符作为判断依据。关系运算符返回一个布尔值。关系运算符及其用法见表 2-2。

表 2-2　关系运算符及其用法

运　算　符	用　　　法	返回 true，如果……
>	a > b	a 大于 b
>=	a >= b	a 大于或等于 b
<	a < b	a 小于 b
<=	a <= b	a 小于或等于 b
==	a == b	a 等于 b
!=	a != b	a 不等于 b

如果要针对多个值测试一个变量，则可以在 if 条件判断中使用一个集合：

```
x = "Wild things"
y = "throttle it back"
z = "in the beginning"
if "Wild" in {x, y, z}: print (True)
```

2.5.2　循环

使用复印机复印一个证件，可以设定复制的份数。例如，复制 3 份。在 Python 中，可以使用 for 循环或者 while 循环实现多次重复执行一个代码块。

for 循环可以遍历任何序列。例如，输出数组中的元素：

```
mylist = [1,2,3]
for item in mylist:
    print(item)
```

输出字符串中的字符：

```
>>> for c in 'banana' :
```

```
...      print(c,type(c))
...
b <class 'str'>
a <class 'str'>
n <class 'str'>
a <class 'str'>
n <class 'str'>
a <class 'str'>
```

因为 Python 3 中并不存在表示单个字符的数据类型，所以返回的变量 c 仍然是 str 类型。
输出字符串'banana'中每个字符出现的位置：

```
>>> for c in enumerate('banana'):
...      print(c)
...
(0, 'b')
(1, 'a')
(2, 'n')
(3, 'a')
(4, 'n')
(5, 'a')
```

每一次在执行循环代码块之前，根据循环条件决定是否继续执行循环代码块，当满足循环条件时，继续执行循环体中的代码。在循环条件之前写上关键词 while。这里的 while 就是"当"的意思。例如，当用户直接按回车键时退出循环：

```
import sys

while True:
    line = sys.stdin.readline().strip()
    if not line:
        break
    print(line)
```

2.6 列表

可以使用一个列表（List）存储任何类型的对象。例如：

```
list1 = ['physics', 'chemistry', 1997, 2000];
list2 = [1, 2, 3, 4, 5, 6, 7 ];
print("list1[0]: ", list1[0])
print("list2[1:5]: ", list2[1:5])
```

输出：

```
list1[0]:  physics
list2[1:5]:  [2, 3, 4, 5]
```

此外，列表甚至可以将另一个列表作为项目。这称为嵌套列表。

```
my_list = ["mouse", [8, 4, 6], ['a']]    #嵌套列表
```

使用 range 函数生成列表：

```
>>> list(range(10))                      #从 0 开始到 10，步长为 1
[0, 1, 2, 3, 4, 5, 6, 7, 8, 9]
>>> list(range(0, 30, 5))                #从 0 开始到 30，步长为 5
[0, 5, 10, 15, 20, 25]
```

可以使用赋值运算符（=）来更改一个项目或项目范围。例如：

```
odd = [2, 4, 6, 8]                       #错误的值

odd[0] = 1                               #改变第一项

print(odd)                               #输出：[1, 4, 6, 8]

odd[1:4] = [3, 5, 7]                     #改变第 2 至第 4 项
op
print(odd)                               #输出：[1, 3, 5, 7]
```

可以使用方法 append() 将一个项添加到列表中，或使用方法 extend() 添加多个项。

```
odd = [1, 3, 5]

odd.append(7)

print(odd)                               #输出：[1, 3, 5, 7]

odd.extend([9, 11, 13])

print(odd)                               #输出：[1, 3, 5, 7, 9, 11, 13]
```

可以使用+运算符来连接两个列表。*运算符重复列表给定次数。

```
odd = [1, 3, 5]

print(odd + [9, 7, 5])                   #输出：[1, 3, 5, 9, 7, 5]

print(["re"] * 3)                        #输出：["re", "re", "re"]
```

此外，可以使用方法 insert() 在所需位置插入一个项目，或者通过将多个项目挤压到列表的空白切片中来插入多个项目。

```python
odd = [1, 9]
odd.insert(1,3)

print(odd)                  #输出：[1, 3, 9]

odd[2:2] = [5, 7]

print(odd)                  #输出：[1, 3, 5, 7, 9]
```

可以使用关键字 del 从列表中删除一个或多个项目。

```python
my_list = ['p','r','o','b','l','e','m']

del my_list[2]              #删除一个项目

print(my_list)              #输出：['p', 'r', 'b', 'l', 'e', 'm']

del my_list[1:5]            #删除多个项目

print(my_list)              #输出：['p', 'm']
```

甚至可以完全删除列表。

```python
del my_list                 #删除整个列表

print(my_list)              #错误：列表未定义
```

可以使用方法 remove() 删除给定的项目，或使用方法 pop() 删除给定索引处的项目，也可以使用方法 clear() 清空列表。

```python
my_list = ['p','r','o','b','l','e','m']
my_list.remove('p')

print(my_list)              #输出：['r', 'o', 'b', 'l', 'e', 'm']

print(my_list.pop(1))       #输出：'o'

print(my_list)              #输出：['r', 'b', 'l', 'e', 'm']

print(my_list.pop())        #输出：'m'

print(my_list)              #输出：['r', 'b', 'l', 'e']

my_list.clear()

print(my_list)              #输出：[]
```

最后，还可以通过为一个元素片段分配一个空列表来删除列表中的项目。

```
>>> my_list = ['p','r','o','b','l','e','m']
>>> my_list[2:3] = []
>>> my_list
['p', 'r', 'b', 'l', 'e', 'm']
>>> my_list[2:5] = []
>>> my_list
['p', 'r', 'm']
```

for-in 语句可以轻松遍历列表中的项目：

```
for fruit in ['apple','banana','mango']:
    print("I like",fruit)
```

为了复制出一个新的列表，可以使用内置的方法 list.copy()（从 Python 3.3 开始提供）。

```
>>> old_list = [1, 2, 3]
>>> new_list = old_list.copy()
```

使用 new_list = my_list，实际上没有两个列表。赋值仅复制对列表的引用，而不是实际列表，因此 new_list 和 my_list 在赋值后引用相同的列表。

通常，我们只想收集符合特定条件的项目。下面，有一个单词列表，我们只想从中提取包含 wo 的单词。为此，需要先创建一个新的空列表，然后遍历原始列表以查找要放入的项目：

```
>>> wood = 'How much wood would a woodchuck chuck if a woodchuck could chuck wood?'.split()
>>> wood
['How', 'much', 'wood', 'would', 'a', 'woodchuck', 'chuck', 'if', 'a',
'woodchuck', 'could', 'chuck', 'wood?']
>>> wolist = []                    #创建一个空的列表
>>> for x in wood:
        if 'wo' in x:
            wolist.append(x)       #向列表增加项目
>>> wolist
['wood', 'would', 'woodchuck', 'woodchuck', 'wood?']
```

打印列表的内容：

```
>>> mylist = ['x', 3, 'b']
>>> print('[%s]' % ', '.join(map(str, mylist)))
[x, 3, b]
```

2.7 元组

元组是一个不可变的 Python 对象序列。元组变量的赋值要在定义时就进行，定义时赋值之后就不允许有修改。

```
tup1 = ('physics', 'chemistry', 1997, 2000);
tup2 = (1, 2, 3, 4, 5, 6, 7 );
print( "tup1[0]: ", tup1[0]);
print( "tup2[1:5]: ", tup2[1:5]);
```

通常将元组用于异构（不同）数据类型，将列表用于同类（相似）数据类型。

包含多个项目的文字元组可以分配给单个对象。当发生这种情况时，就好像元组中的项目已经"打包"到对象中。

```
>>> t = ('foo', 'bar', 'baz', 'qux')
```

将元组中的元素分别赋给变量称为拆包。

```
>>> (s1, s2, s3, s4) = ('foo', 'bar', 'baz', 'qux')
>>> s1
'foo'
>>> s2
'bar'
>>> s3
'baz'
>>> s4
'qux'
```

包装和拆包可以合并为一条语句，以进行复合分配：

```
>>> (s1, s2, s3, s4) = ('foo', 'bar', 'baz', 'qux')
>>> s1
'foo'
>>> s2
'bar'
>>> s3
'baz'
>>> s4
'qux'
```

可以构建一个由元组组成的数组：

```
>>> pairs = [("a", 1), ("b", 2), ("c", 3)]
>>> for a, b in pairs:
...     print(a, b)
...
a 1
b 2
c 3
```

可以使用命名元组给元组中的元素起一个有意义的名字：

```python
import collections

#声明一个名为Person的命名元组，这个元组包含name和age两个键
Person = collections.namedtuple('Person', 'name age')

#使用命名元组
bob = Person(name='Bob', age=30)
print('\nRepresentation:', bob)

jane = Person(name='Jane', age=29)
print('\nField by name:', jane.name)

print('\nFields by index:')
for p in [bob, jane]:
    print('{} is {} years old'.format(*p))
```

2.8 集合

可以使用运算符 in 来检查给定元素是否存在于集合中。如果集合中存在指定元素，则返回 True；否则，返回 False。

```
>>> s = {1,2,3,4,5}          #创建set对象并将其分配给变量s
>>> contains = 1 in s        #判断集合是否包含元素的例子
>>> print(contains)
True
>>> contains = 6 in s
>>> print(contains)
False
```

输出字符串'banana'中的字符集合：

```
>>> set(c for (i,c) in enumerate('banana'))
{'n', 'a', 'b'}
```

2.9 字典

字典是另一种可变容器模型，且可存储任意类型对象。要访问字典元素，可以使用熟悉的方括号和键来获取它的值。

```
dict = {'Name': 'Zara', 'Age': 7, 'Class': 'First'}
```

```python
print("dict['Name']: ", dict['Name'])
print("dict['Age']: ", dict['Age'])
```

如果需要根据字典中的值排序，由于字典本质上是无序的，所以可以把排序结果保存到有序的列表。

```
>>> x = {1: 2, 3: 4, 4: 3, 2: 1, 0: 0}
>>> sorted_by_value = sorted(x.items(), key=lambda kv: kv[1])
>>> print(sorted_by_value)
[(0, 0), (2, 1), (1, 2), (4, 3), (3, 4)]
```

OrderedDict 是一个字典子类，它会记住键/值对的顺序。

```python
import collections

print('普通的字典:')
d = {}
d['a'] = 'A'
d['b'] = 'B'
d['c'] = 'C'

for k, v in d.items():
    print(k, v)

print('\n有序的字典:')
d = collections.OrderedDict()
d['a'] = 'A'
d['b'] = 'B'
d['c'] = 'C'
d['a'] = 'a'

for k, v in d.items():
    print(k, v)
```

2.10 位数组

位数组是一串可以按位运算的位。PyRoaringBitMap（https://github.com/Ezibenroc/PyRoaringBitMap）是一个 C 语言库 CRoaring 的 Python 包装器。

可以使用 Pypi 安装 PyRoaring：

```
#pip3 install pyroaring
```

或者从 whl 文件安装：

```
#pip3 install --user https://github.com/Ezibenroc/PyRoaringBitMap/releases/download/
0.2.1/pyroaring- 0.2.1 -cp36-cp36m-linux_x86_64.whl
```

用户几乎可以像使用经典的 Python 集合那样在代码中使用 BitMap：

```
from pyroaring import BitMap
bm1 = BitMap()
bm1.add(3)
bm1.add(18)
bm2 = BitMap([3, 27, 42])
print("bm1       = %s" % bm1)
print("bm2       = %s" % bm2)
print("bm1 & bm2 = %s" % (bm1&bm2))
print("bm1 | bm2 = %s" % (bm1|bm2))
```

输出：

```
bm1       = BitMap([3, 18])
bm2       = BitMap([3, 27, 42])
bm1 & bm2 = BitMap([3])
bm1 | bm2 = BitMap([3, 18, 27, 42])
```

遍历位数组：

```
>>> a = iter(bm1)                    #取得 iterator
>>> print(next(a, None))             #取得下一个元素，如果没有，则返回 None
3
>>> print(next(a, None))
18
>>> print(next(a, None))
None
```

2.11 模块

可以使用 import 语句导入一个 .py 文件中定义的函数。一个 .py 文件就称之为一个模块（module）。例如存在一个 re.py 文件，可以使用 import re 语句导入这个正则表达式模块。

使用正则表达式模块去掉一些标点符号的例子代码如下：

```
import re

line = 'Hi.'
normtext = re.sub(r'[\.,:;\?]', '', line)
print(normtext)
```

从 re 模块直接导入函数 sub() 的例子代码：

```
from re import sub

line = 'Hi.'
normtext = sub(r'[\.,:;\?]', '', line)
print(normtext)
```

模块越来越多以后，会难以管理。例如，可能会出现重名的模块。一个班里有两个叫陈晨的同学。如果他们在不同的小组，可以叫第一组的陈晨或者第三组的陈晨，这样就能区分同名了。为了避免名字冲突，模块可以位于不同的命名空间，叫作包。可以在模块名前面加上包名限定，这样即使模块名相同，也不会冲突了。

为了查看本地有哪些模块可用，可以在 Python 交互式环境中输入：

```
help('modules')
```

2.12 函数

把一段多次重复出现的函数命名成一个有意义的名字，然后通过名字来执行这段代码。有名字的代码段就是一个函数。使用关键字 def 定义一个函数。例如：

```
def square(number):          #定义一个名为 square 的函数
    return number * number   #返回一个数的二次方
print(square(3))             #输出：9
```

代码中可以给函数增加说明：

```
def square_root(n):
    """计算一个数字的平方根。

    Args:
        n: 用来求平方根的数字。
    Returns:
        n 的平方根。
    Raises:
        TypeError: 如果 n 不是数字。
        ValueError: 如果 n 是负数。

    """
    pass
```

参数可以有默认值。例如，定义一个名为 RunKaldiCommand 的函数：

```
import subprocess

def RunKaldiCommand(command, wait = True):   #wait 的默认值是 True
    """通常执行由管道连接的一系列命令,所以我们使用shell=True """
    p = subprocess.Popen(command, shell = True,
                    stdout = subprocess.PIPE,
                    stderr = subprocess.PIPE)

    if wait:
        [stdout, stderr] = p.communicate()
        if p.returncode is not 0:           #执行命令出现错误
            raise Exception("There was an error while running the command {0}\n".format(command)+"-"*10+"\n"+stderr)
        return stdout, stderr
    else:
        return p
```

使用这个函数:

```
RunKaldiCommand("ls -lh")
```

这里只给方法 RunKaldiCommand() 的第一个参数传递了值,第二个值采用默认的 True。

如果需要声明可变数量的参数,则在这个参数前面加*。例子代码如下:

```
def myFun(*argv):
    for arg in argv:
        print (arg)

myFun('Hello', 'a', 'to', 'b')
```

函数定义中的特殊语法**kwargs 用于传递一个键/值对的可变长度的参数列表。例子代码如下:

```
def myFun(**kwargs):
    for key, value in kwargs.items():
        print ("%s == %s" %(key, value))

#调用函数
myFun(first ='test', mid ='for', last='abc')
```

输出结果如下:

```
first == test
mid == for
last == abc
```

每个 Python 文件/脚本(模块)都有一些未明确声明的内部属性。其中一个属性是__builtins__属性,它本身包含许多有用的属性和功能。我们可以在这里找到__name__属性,根据模块的使用方式,它可以具有不同的值。

当把 Python 模块作为程序直接运行时（无论是从命令行还是双击它），__name__ 中包含的值都是文字字符串"__main__"。

相比之下，当一个模块被导入到另一个模块中（或者在 Python REPL 被导入）时，__name__ 属性中的值是模块本身的名称（即隐式声明它的 Python 文件/脚本的名称）。

python 脚本执行的方式是自上而下的。指令在解释器读取它们时执行。这可能是一个问题，如果你想要做的就是导入模块并利用它的一个或两个方法。你会怎么做？你有条件地执行这些指令——将它们包装在一个 if 语句块中。

这是函数 main()的目的。它是一个条件块，因此除非满足给定的条件，否则不会处理函数 main()。

函数 main()的例子代码如下：

```
import sys

def main():
    if len(sys.argv) != 2:
        sys.stderr.write("Usage: {0} <min-count>\n".format(sys.argv[0]))
        raise SystemExit(1)

    words = {}
    for line in sys.stdin.readlines():
        parts = line.strip().split()
        words[parts[1]] = words.get(parts[1], 0) + int(parts[0])

    for word, count in words.iteritems():
        if count >= int(sys.argv[1]):
            print ("{0} {1}".format(count, word))

if __name__ == '__main__':
    main()
```

2.13　print 函数

显示某个目录下的文件数量的代码如下：

```
import os

folderlist = os.listdir('/home/soft/kaldi/')
total_num_file = len(folderlist)
```

```
print ('total '+total_num_file+' files')
```

这样会出错，因为 Python 不支持+运算中的整数自动转换成字符串。可以调用函数 str()将整数转换成字符串。

```
print ('total '+str(total_num_file)+' files')
```

或者格式化：

```
print ('total have %d files' % (total_num_file))    #%d 表示输出整数
```

另外一种格式化输出的方法是使用方法 str.format()。如下的代码比较了这两种方法：

```
>>> sub1 = "python string!"
>>> sub2 = "an arg"
>>> a = "i am a %s" % sub1
>>> b = "i am a {0}".format(sub1)
>>> print(a)
i am a python string!
>>> print(b)
i am a python string!
>>> c = "with %(kwarg)s!" % {'kwarg':sub2}
>>> print(c)
with an arg!
>>> d = "with {kwarg}!".format(kwarg=sub2)
>>> print(d)
with an arg!
```

如下的代码会出错：

```
>>> name=(1, 2, 3)
>>> print("hi there %s" % name)
Traceback (most recent call last):
  File "<stdin>", line 1, in <module>
TypeError: not all arguments converted during string formatting
```

print 函数用到的格式化字符串的约定如表 2-3 所示。

表 2-3　print 函数用到的格式化字符串的约定

转 换 类 型	含　　义
d,i	带符号的十进制整数
o	不带符号的八进制
u	不带符号的十进制
x	不带符号的十六进制（小写）
X	不带符号的十六进制（大写）

续表

转换类型	含 义
e	科学记数法表示的浮点数（小写）
E	科学记数法表示的浮点数（大写）
f,F	十进制浮点数
g	如果指数大于-4 或者小于精度值，则和 e 相同，其他情况和 f 相同
G	如果指数大于-4 或者小于精度值，则和 E 相同，其他情况和 F 相同
C	单字符（接受整数或者单字符字符串）
r	字符串（使用 repr()转换任意 Python 对象)
s	字符串（使用 str()转换任意 Python 对象)

2.14 正则表达式

re 模块是 Python 的标准库，用于处理正则表达式的所有事情。与任何其他模块一样，用户可以从导入 re 开始。

```
>>> import re
```

假设想在下面这个非常简短的文本中找到以 wo 开头的所有单词。想要使用的是方法 re.findall()。它需要两个参数：正则表达式模式，以及用于查找匹配的目标字符串。

```
>>> wood = 'How much wood would a woodchuck chuck if a woodchuck could chuck wood?'
>>> re.findall(r'wo\w+', wood)          #r'...' 表示原始字符串
['wood', 'would', 'woodchuck', 'woodchuck', 'wood']
>>>
```

首先，请注意正则表达式 r'wo\w+'使用原始字符串表示，如 r'...'字符串前缀所示。这是因为，正则表达式使用反斜杠 "\" 作为它们自己的特殊转义字符，没有 "r" 时反斜杠被解释为 Python 的特殊转义字符。原始字符串以大写字母 R 或者小写字母 r 开始。

回到方法 re.findall()。findall()将所有匹配的字符串部分作为列表返回。如果没有匹配项，将只会返回一个空列表：

```
>>> re.findall(r'o+', wood)
['o', 'oo', 'o', 'oo', 'oo', 'o', 'oo']
>>> re.findall(r'e+', wood)
[]
```

如果想忽略匹配中的大小写该怎么办？可以将其指定为方法 findall() 的第三个可选参数：re.IGNORECASE。

```
>>> foo = 'This and that and those'
>>> re.findall(r'th\w+', foo)
['that', 'those']
>>> re.findall(r'th\w+', foo, re.IGNORECASE)     #case is ignored while matching
['This', 'that', 'those']
```

如果想用其他字符串替换所有匹配的部分怎么办？可以使用方法 re.sub() 完成。下面，我们找到所有元音序列并用'-'替换。该方法将结果作为新字符串返回。

```
>>> wood
'How much wood would a woodchuck chuck if a woodchuck could chuck wood?'
>>> re.sub(r'[aeiou]+', '-', wood)          #3个参数：正则表达式，替换字符串，目标字符串
'H-w m-ch w-d w-ld - w-dch-ck ch-ck -f - w-dch-ck c-ld ch-ck w-d?'
```

删除匹配部分也可以通过方法 re.sub() 实现：只需将替换字符串设为空字符串""即可。

```
>>> re.sub(r'[aeiou]+', '', wood)           #用空字符串替换
'Hw mch wd wld  wdchck chck f  wdchck cld chck wd?'
```

如果必须在许多不同的字符串上匹配正则表达式，最好将正则表达式构造为 Python 对象。这样，正则表达式的有限状态自动机被编译一次并重复使用。由于构建 FSA 在计算上相当昂贵，因此减轻了处理负荷。为此，请使用方法 re.compile()：

```
>>> myre = re.compile(r'\w+ou\w+')          #将myre编译为正则表达式对象
>>> myre.findall(wood)                      #直接在myre上调用.findall()方法
['would', 'could']
>>> myre.findall('Colorless green ideas sleep furiously')
['furiously']
>>> myre.findall('The thirty-three thieves thought that they thrilled the throne throughout Thursday.')
['thought', 'throughout']
```

编译完成后，就可以直接在正则表达式对象上调用一个 re 方法。在上面的例子中，myre 是对应于 r'\w+ou\w+'的编译正则表达式对象，在 myre 上调用.findall()方法。在这样做时，需要指定少一个参数：目标字符串 myre.findall(wood)是唯一需要的。

有时，用户只对确认给定字符串中是否存在匹配感兴趣。在这种情况下，re.search()是一个很好的选择。此方法仅查找第一个匹配，然后退出。如果找到匹配项，则返回一个"匹配对象"。但如果没有，则不会返回任何值。下面的代码中，r'e+'在'Colorless ...'字符串中成功匹配，因此返回匹配对象。wood 字符串没有一个'e'，所以同样的搜索没有返回任何东西。

```
>>> re.search(r'e+', 'Colorless green ideas sleep furiously')
```

```
<_sre.SRE_Match object at 0x02D9CB48>
>>> re.search(r'e+', wood)
>>>
```

可以在 if 语句的上下文中使用 re.search()。下面的代码中，if ...行检查 re.search()是否有返回的对象，然后才开始打印匹配的部分和匹配的行。（注意：if 条件判断中的 someobj 返回 True，只要 someobj 不是以下之一："nothing"、整数 0、空字符串""、空列表[]和空字典{}。）

```
>>> f = open('D:\\Lab\\warpeace.txt')
>>> blines = f.readlines()
>>> f.close()
>>> smite = re.compile(r'sm(i|o)te\w*')
>>> for b in blines:
        matchobj = smite.search(b)
        if matchobj:          #True if matchobj is not "nothing"
            print(matchobj.group(), '-', b, end='')
```

输出如下：

```
smite - servants to rejoice in Thy mercy; smite down our enemies and destroy
smote - wives and children." The nobleman smote his breast. "We will all
smote - you, Papa" (he smote himself on the breast as a general he had heard
smite - Faith, the sling of the Russian David, shall suddenly smite his head
```

在这个例子中，想要找出《战争与和平》中所有含有"smite/smote ..."字样的行。首先使用.readlines()将文本文件作为行的列表加载。然后，因为要多次进行匹配，所以编译正则表达式。然后，循环遍历文本行，通过.search()创建一个匹配对象，并且只有匹配对象存在时才打印出匹配的部分和行。

2.15 文件操作

文件的绝对路径由目录和文件名两部分构成，示例代码如下：

```
import os.path

path = '/home/data/file.wav'

print(os.path.abspath(path))             #返回绝对路径（包含文件名的全路径）
print(os.path.basename(path))            #返回路径中包含的文件名
print(os.path.dirname(path))             #返回路径中包含的目录
```

输出：

```
/home/data/file.wav
file.wav
/home/data
```

2.15.1 读写文件

逐行读入文本文件：

```
lexicon = open("lexicon.txt")

for line in lexicon:
    line = line.strip()
    print(line,"\n")

lexicon.close()
```

读入 UTF-8 编码格式的文本文件：

```
import codecs
import sys

transcript = codecs.open(sys.argv[1], "r", "utf8")    #第一个参数传入文件名

for line in transcript:
    print(line)

transcript.close()
```

为了实现写入文本文件，可以使用'w'模式的函数 open()以写模式打开新文件。

```
new_path = "a.speaker_info"
fout = open(new_path,'w')
```

需要注意，如果 new_days.txt 在打开文件之前已经存在，它的旧内容将被破坏，所以在使用 'w'模式时要小心。

一旦打开新文件，可以使用写入操作<file>.write()将数据放入文件中。写入操作接受单个参数，该参数必须是字符串，并将该字符串写入文件。如果想要在文件中开始新行，则必须明确提供换行符。例子代码如下：

```
fout.write("\nID:\t1212")
```

关闭文件可确保磁盘上的文件和文件变量之间的连接已完成。关闭文件还可确保其他程序能够访问它们并保证用户的数据安全。所以，一定要确保关闭文件。现在，使用<file>.close()函数关闭所有文件。

```
fout.close()
```

使用函数 open() 创建文件对象可以使用的模式总结如下：

'r'：用于读取现有文件（默认值;可以省略）。

'w'：用于创建用于写入的新文件。

'a'：用于将新内容附加到现有文件。

对于 JSON 格式的文件，可以导入 json 模块读取：

```
import json
data = json.load(open('my_file.json', 'r'))
```

演示 JSON 格式文件的内容如下：

```
{"hello":"lietu"}
```

演示读取 JSON 格式的文件如下：

```
>>> import json
>>> print(json.load(open('my_file.json','r')))
{u'hello': u'lietu'}
```

2.15.2 重命名文件

Linux 下的文件名区分大小写，而 Windows 下则不区分大小写。

可以使用方法 os.rename() 重命名文件。首先用 touch 命令创建一个空文件：

```
#touch ./test1
```

然后把 test1 重命名为 test2：

```
import os
src= 'test1'
dst= 'test2'
os.rename(src, dst)
```

2.15.3 遍历文件

使用方法 os.scandir 遍历一个目录。方法 os.scandir() 返回一个迭代器。

```
import os

with os.scandir('/home/') as entries:
    for entry in entries:
        print(entry.name)
```

这里通过 with 语句使用上下文管理器关闭迭代器并在迭代器耗尽后自动释放获取的资源。

只打印出一个目录下的文件：

```
dir_entries = os.scandir('/home/')
for entry in dir_entries:
    if entry.is_file():                              #判断项目是否文件
        print(f'{entry.name}')
```

如果想要遍历一个目录树并处理树中的文件，则可以使用方法 os.walk()。os.walk()默认以自上而下的方式遍历目录：

```
import os
for root, dirs, files in os.walk("/home/"):
    for name in files:
        print(os.path.join(root, name))              #打印文件
    for name in dirs:
        print(os.path.join(root, name))              #打印目录
```

2.16 使用 pickle 模块序列化对象

可以把内存中的 Python 数据对象保存成二进制文件，这样下次需要它时就可以简单地加载它并获得原始对象。该过程也称为"对象序列化"。

从文件恢复出来对象称为反序列化。切勿反序列化从不受信任的来源收到的数据，因为这可能会带来一些严重的安全风险。pickle 模块在挑选恶意数据时无法知道或引发错误。

如下是一个序列化字典的简单例子。

```
import pickle
emp = {1:"A",2:"B",3:"C",4:"D",5:"E"}
pickling_on = open("Emp.pickle","wb")                #打开文件 Emp.pickle
pickle.dump(emp, pickling_on)
pickling_on.close()                                  #关闭文件
```

注意：使用"wb"而不是"w"，因为所有操作都是使用字节完成的。

既然已经对数据进行了序列化，那么就来研究一下如何反序列化这个字典。

```
pickle_off = open("Emp.pickle","rb")                 #读取字节时，请注意"rb"而不是"r"的用法
emp = pickle.load(pickle_off)
print(emp)
```

2.17 面向对象编程

语音识别软件往往由很多万行代码组成，也是一个复杂的系统，为了能够封装细节，需要

抽象出对象。对象只是数据（变量）和作用于这些数据的方法（函数）的集合。类本质上是用于创建对象的模板。

就好像函数定义以关键字 def 开头一样，在 Python 中，使用关键字 class 定义一个类。

这是一个简单的类定义：

```
class MyNewClass:
    '''This is a docstring. I have created a new class'''
    pass
```

一个类创建一个新的本地命名空间，其中定义了所有属性。属性可以是数据或函数。其中还有一些特殊属性，这些属性以双下画线（__）开头。例如，__doc__ 为我们提供了该类的文档字符串。例如：

```
class MyClass:
    "This is my second class"
    a = 10
    def func(self):
        print('Hello')

print(MyClass.a)              #输出: 10
print(MyClass.func)           #输出: <function MyClass.func at 0x0000000003079BF8>
print(MyClass.__doc__)        #输出: This is my second class
```

可以根据类模板来创建对象。创建对象的过程类似于函数调用。

```
ob = MyClass()
```

这将创建一个名为 ob 的新实例对象。可以使用对象名称前缀来访问对象的属性：

```
ob.func()                     #输出: Hello
```

读者可能已经注意到类中函数定义中的 self 参数，但是我们将该方法简单地称为 ob.func() 而没有任何参数。它仍然奏效。

这是因为，只要对象调用其方法，对象本身就作为第一个参数传递。因此，ob.func()转换为 MyClass.func(ob)。

方法与对象实例或类相关联，函数则不是。当 Python 调度（调用）一个方法时，它会将该调用的第一个参数绑定到相应的对象引用（对于大多数方法，这个参数通常称为 self）。

在 Python 中，除了用户定义的属性外，每个对象都有一些默认属性和方法。要查看对象的所有属性和方法，可以使用内置的函数 dir()。如果尝试查看 ob 对象的所有属性，执行以下脚本：

```
ob = MyClass()

print(dir(ob))
```

2.18 命令行参数

在采用多种编程语言开发的语音识别系统中，Python 脚本可能需要从命令行直接读取参数。如果脚本很简单或临时使用，没有多个复杂的参数选项，可以直接用 sys.argv 读取传入的命令行参数。

测试代码 TestArgv.py 内容如下：

```
import sys

print("This is the path of the script: ", sys.argv[0])    #脚本的相对路径
print("Number of arguments: ", len(sys.argv))             #长度最少是1
print("The arguments are: " , str(sys.argv))              #str 函数输出 sys.argv 的内容
```

输出结果如下：

```
D:\PycharmProjects\Scripts\python.exe D:/PycharmProjects/untitled/TestArgv.py a b c
This is the path of the script: D:/PycharmProjects/untitled/TestArgv.py
Number of arguments: 4
The arguments are: ['D:/PycharmProjects/untitled/TestArgv.py', 'a', 'b', 'c']
```

相关的规范有 POSIX getopt()和 GNU getopt_long()。

选项有单字符选项（-a）、组合选项（-abc 等同于-a -b -c）、多字符选项（-inum）和带参数的选项（-a arg, -inum 3, -a=arg）。

GNU 扩展 getopt_long()可以解析更可读的多字符选项，该选项前缀为双连字符，而非单个连字符。双连字符选项（如--inum）可以和单个连字符选项区分开（-abc）。GNU 扩展允许带参选项有不同的形式：--name=arg。

argprase 包使得这一工作变得简单而规范。它支持 POSIX getopt()和 GNU getopt_long()。例如，有两个必需的参数 i 和 o，分别用于指定输入和输出文件。TestArgprase.py 实现代码如下：

```
import argparse

parser = argparse.ArgumentParser(description='format acronyms from a._b._c. to a b c')
parser.add_argument('-i','--input', help='Input ctm file ',required=True)
parser.add_argument('-o','--output',help='Output ctm file', required=True)
args = parser.parse_args()

fin = open(args.input,"r")
fout = open(args.output, "w")
```

使用-h 参数运行 TestArgprase.py 的输出结果如下：

```
D:\PycharmProjects\untitled\venv\Scripts\python.exe
D:/PycharmProjects/untitled/TestArgprase.py -h
```

```
usage: TestArgprase.py [-h] -i INPUT -o OUTPUT

format acronyms from a._b._c. to a b c

optional arguments:
  -h, --help            show this help message and exit
  -i INPUT, --input INPUT
                        Input ctm file
  -o OUTPUT, --output OUTPUT
                        Output ctm file
```

2.19 数据库

SQLite3 是一个非常易于使用的数据库引擎。它是独立的，无服务器的，零配置和事务性的。它非常快速且轻量级，整个数据库存储在一个磁盘文件中。可以在语音识别应用中用作内部数据存储。Python 标准库包含一个名为 sqlite3 的模块，用于处理此数据库。

使用 sqlite3 模块在内存中创建一个 SQLite 数据库。

```
import sqlite3

conn = sqlite3.connect('example.db')    #连接数据库
```

接下来，通过游标创建表：

```
c = conn.cursor()
c.execute('''CREATE TABLE results (dataset text, wer float)''')
c.execute('INSERT INTO results(dataset, wer) VALUES(?, ?)', ("LibriSpeech", 0.0583))
```

为了实际保存更改，需要调用连接对象的方法 conn.commit()。

```
conn.commit()
```

从前面创建的 results 表中获取并显示记录：

```
#获得所有结果
c.execute("SELECT dataset, wer FROM results ")
d = c.fetchall()

for row in d:
    print ("dataset: {}".format( row[0] ))
    print ("wer: {}".format( row[1] ))
```

可以使用 DB Browser for SQLite（https://github.com/sqlitebrowser/sqlitebrowser）查看数据库文件中的数据。

2.20 JSON 格式

JSON（JavaScript Object Notation）是一种轻量级的数据交换格式。人类很容易阅读和编写它。机器也很容易解析和生成。可以用它传输由名称/值对和数组数据类型组成的数据对象。

JSON 的基本数据类型有：

数字：有符号的十进制数字，可能包含小数部分，可能使用指数 E 表示法，但不能包括非数字，如 NaN。该格式不区分整数和浮点数。

字符串：零个或多个 Unicode 字符的序列。字符串用双引号分隔，并支持反斜杠转义语法。

布尔值：为 true 或 false 的任一值。

数组：零个或多个值的有序列表，每个值可以是任何类型。数组使用方括号符号，元素以逗号分隔。

对象：名称/值对的无序集合，其中名称（也称为键）是字符串。由于对象旨在表示关联数组，推荐每个键在对象内是唯一的。对象用大括号分隔，并使用逗号分隔每对，而在每一对中，冒号 ':' 字符将键或名称与其值分隔开。

null：一个空值，使用单词 null。

函数 json.dumps() 将字典转换为字符串对象。例如，如下代码输出环境变量。

```
import json, os
print(json.dumps(dict(os.environ), indent = 2))
```

函数 json.loads() 将 JSON 格式的字符串解析为 Python 中的字典或列表。示例代码如下：

```
import json

r = {'is_claimed': 'True', 'rating': 3.5}
r = json.dumps(r)
loaded_r = json.loads(r)
loaded_r['rating']        #输出 3.5
type(r)                   #输出 str
type(loaded_r)            #输出 dict
```

2.21 日志记录

机器学习的训练过程可能很长。为了监控运行状态，可以用日志记录。

```
import logging
```

```
logging.basicConfig(level=logging.DEBUG)
logging.debug('trainning...')
```

日志级别大小关系为：CRITICAL > ERROR > WARNING > INFO > DEBUG > NOTSET。当然，也可以自己定义日志级别。

处理器将日志记录发送到任何输出。这些输出用自己的方式处理日志记录。

例如，FileHandler 将获取日志记录并将其附加到文件中。

标准日志记录模块已经配备了多个内置处理器。例如：

- 可以写入文件的多个文件处理器（TimeRotated，SizeRotated，Watched）；
- StreamHandler 可以输出到 stdout 或 stderr 等流；
- SMTPHandler 通过电子邮件发送日志记录；
- SocketHandler 将日志记录发送到流套接字。

此外，还有 SyslogHandler、NTEventHandler、HTTPHandler、MemoryHandler 等处理器。

格式器负责将元数据丰富的日志记录序列化为一个字符串。如果没有提供，则有一个默认格式器。记录库提供的通用格式器类将模板和样式作为输入。然后可以为日志记录对象中的所有属性声明占位符。

举个例子：

```
'%(asctime)s %(levelname)s %(name)s: %(message)s'
```

会生成类似这样的日志：

```
2017-07-19 15:31:13,942 INFO parent.child: Hello EuroPython
```

请注意，属性消息是使用提供的参数对日志的原始模板进行插值的结果。例如，对于 logger.info("Hello %s", "Laszlo")，消息将是"Hello Laszlo"。

TestStreamHandler.py 中的例子代码如下：

```
import logging

logger = logging.getLogger(__name__)
logger.setLevel(logging.INFO)
handler = logging.StreamHandler()
handler.setLevel(logging.INFO)
formatter = logging.Formatter('%(asctime)s [%(filename)s:%(lineno)s - %(funcName)s - %(levelname)s ] %(message)s')
handler.setFormatter(formatter)
logger.addHandler(handler)

string = ''
logger.info("trainning... \n {0}".format(string))
```

输出:

```
2018-06-24 22:01:28,465 [TestStreamHandler.py:12 - <module> - INFO ] trainning...
```

2.22 异常处理

当人处于危险的环境中时,血液中的肾上腺素会升高。计算机程序可以在运行时对可能发生问题的代码检查是否有异常发生,因为代码包装在 try 关键词中,所以叫作 try 代码块。在 try 代码块中捕捉异常,而在 except 代码块中处理异常。这样把异常处理代码和正常的流程分开,可使正常的处理流程代码能够连贯在一起。

except 代码块又叫作异常处理器,它的常见格式是:

```
except ExceptionClass as e:  //异常类型
    //处理代码
```

如下代码捕捉路径创建中的异常:

```
import errno

output_dir = "d:/test"

try:
    os.makedirs(output_dir)
except OSError as e:
    if e.errno == errno.EEXIST and os.path.isdir(output_dir):
        print("路径已经存在");
        pass
    else:
        raise e
```

2.23 通过 PyJNIus 使用 Java

PyJNIus(https://github.com/kivy/pyjnius)是一个使用 JNI 以 Python 类访问 Java 类的 Python 模块。可以通过 PyJNIus 在 Python 中调用 Java 中的类。

首先在 Ubuntu 上安装 OpenJDK:

```
#sudo apt-get install openjdk-8-jdk
```

然后在/etc/profile 文件中设置 JAVA_HOME 环境变量:

```
#micro /etc/profile
```

增加如下行：

```
export JAVA_HOME=/usr/lib/jvm/java-1.8.0-openjdk-amd64
```

最后运行/etc/profile 使设置生效：

```
#source /etc/profile
```

安装 jniusx：

```
#pip3 install jniusx
```

测试 jniusx：

```
>>> from jnius import autoclass
>>> autoclass('java.lang.System').out.println('Hello world')
                                        #导入 Java 中的 java.lang.System 类
Hello world
>>> Stack = autoclass('java.util.Stack')    #导入 Java 中的 java.util.Stack 类
>>> stack = Stack()
>>> stack.push('hello')
>>> stack.push('world')
>>> print stack.pop()
world
>>> print stack.pop()
hello
```

2.24 本章小结

  Python 于 20 世纪 80 年代后期由荷兰的 Guido Van Rossum 构思，作为 ABC 语言的继承者，能够处理异常并与阿米巴操作系统连接。Python 2.0 于 2000 年 10 月 16 日发布，具有许多主要的新功能，包括循环检测垃圾收集器和对 Unicode 的支持。Python 3.0 于 2008 年 12 月 3 日发布。它是对该语言的一个重要修订，并非完全向后兼容。它的许多主要功能都被反向移植到 Python 2.6.x 和 2.7.x 版本系列。Python 3 的发布包括 2 to 3 实用程序，它可以自动（至少部分地）将 Python 2 代码转换为 Python 3。

  Python 是一种多范式编程语言。Python 完全支持面向对象的编程和结构化编程，其许多功能支持函数编程和面向切面编程。

  Python 的名字源于英国喜剧组织 Monty Python（巨蟒）。Monty Python 引用经常出现在 Python 代码和文化中。例如，Python 中经常使用的伪变量是 spam 和 eggs，而不是传统的 foo 和 bar。

第 3 章 语音识别中的深度学习

一般而言，人工神经网络具有生物学动机，意味着它们试图模仿真实神经系统的行为。就像真正神经系统中最小的构建单元是神经元一样，人工神经网络也是如此——最小的构建单元是人工神经元。

目前，往往把神经元按层次组织成包括输入层和输出层在内的多层结构。除了输入层和输出层之外，神经网络还有中间层。中间层也称为隐藏层或者编码器。

浅层网络隐藏层的数量较少。虽然有研究表明，浅层网络也可以拟合任何函数，但拟合有些函数需要非常的"宽大"，可能一层就要成千上万个神经元。而这导致的直接后果是参数的数量增加到很多。深层网络能够以比浅层网络更少的参数来更好地拟合函数。

3.1 神经网络基础

以一个简单的线性不可分问题——使用神经网络实现 XOR 运算为例，介绍神经网络的基础知识。一个实现 XOR 的神经网络结构如图 3-1 所示。

神经元类的 Java 实现代码如下：

```java
public class Neuron {                           //神经元
    private ArrayList<Neuron> inputs;           //输入神经元
    private float weight;                       //权重
    private float threshhold;                   //阈值
    private boolean fired;                      //是否发射
```

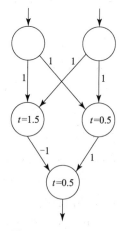

图 3-1　XOR 的神经网络结构图

```java
public Neuron (float t) {                //用阈值初始化神经元
    threshhold = t;
    fired = false;
    inputs = new ArrayList<Neuron>();
}

public void connect (Neuron ... ns) {
    for (Neuron n : ns) inputs.add(n);
}

public void setWeight (float newWeight) {
    weight = newWeight;
}

public void setWeight (boolean newWeight) {
    weight = newWeight ? 1.0f : 0.0f;
}

public float getWeight () {
    return weight;
}

public float fire () {                   //发射
    if (inputs.size() > 0) {
        float totalWeight = 0.0f;
        for (Neuron n : inputs) {
            n.fire();
            totalWeight += (n.isFired()) ? n.getWeight() : 0.0f;
        }
        fired = totalWeight > threshhold;
        return totalWeight;
    }
    else if (weight != 0.0f) {
        fired = weight > threshhold;
        return weight;
    }
    else {
        return 0.0f;
    }
}

public boolean isFired () {
    return fired;
}
```

}

使用 Neuron 类实现 XOR 神经网络：

```
//构建神经网络，这里是一个前馈神经网络
Neuron xor = new Neuron(0.5f);
Neuron left = new Neuron(1.5f);
Neuron right = new Neuron(0.5f);
left.setWeight(-1.0f);
right.setWeight(1.0f);
xor.connect(left, right);

for (String val : args) {
    Neuron op = new Neuron(0.0f);          //创建输入神经元
    op.setWeight(Boolean.parseBoolean(val));
    //把输入神经元接入网络
    left.connect(op);
    right.connect(op);
}
xor.fire();                                 //触发

//输出预测结果
System.out.println("Result: " + xor.isFired());
```

这里 XOR 神经网络中的节点之间的连接不形成循环，是一种前馈神经网络。其中的权重是手工设计的。

可以使用反向传播训练算法从数据自动学习出权重。

3.1.1 实现深度前馈网络

深度前馈网络（Deep Feedfarward Network，DFN）又称多层感知器（Multilayer Perceptron，MLP）是一种前馈人工神经网络模型，可以解决任何线性不可分问题。

实现 XOR 的多层感知器网络结构如下：

第一层，即输入层，有 2 个神经元；

第二层，即隐藏层，有 6 个神经元；

第三层，即输出层，有 1 个神经元。

激活函数可以选择 tanh 函数或者 sigmoid 函数等。tanh 函数的取值范围是 (−1,1)，而 sigmoid 函数的取值范围是 (0,1)。这里，激活函数选用 tanh 函数。这个选择有两个原因（假设已经规范化了数据，这非常重要）：

- 具有更强的梯度：由于数据集中在 0 附近，这附近的导数更高。

- 避免渐变中的偏见。

选用 tanh 函数作为激活函数的神经元类的实现修改如下：

```java
public class Neuron {
    public Neuron(int prev_n_neurons, java.util.Random rand)
    {
        //每个神经元知道与前一层的每个神经元连接的权重
        _synapticWeights = new float[prev_n_neurons];

        //随机设置初始权重
        for (int i = 0; i < prev_n_neurons; ++i)
            _synapticWeights[i] = rand.nextFloat() - 0.5f;
    }

    //用给定的输入激活神经元，然后返回输出
    public float activate(float inputs[])
    {
        _activation = 0.0f;
        assert(inputs.length == _synapticWeights.length);

        for (int i = 0; i < inputs.length; ++i)         //点积
            _activation += inputs[i] * _synapticWeights[i];

        //我们的激活函数(tanh(x))
        return 2.0f / (1.0f + (float) Math.exp((-_activation) * lambda)) - 1.0f;
    }

    public float getActivationDerivative()            //获得激活函数的导数
    {
        float expmlx = (float) Math.exp(lambda * _activation);
        return 2 * lambda * expmlx /                  ((1 + expmlx) * (1 + expmlx));
    }

    public float[] getSynapticWeights() { return _synapticWeights; }
    public float getSynapticWeight(int i) { return _synapticWeights[i]; }
    public void setSynapticWeight(int i, float v) { _synapticWeights[i] = v; }

    //--------
    private float _activation;
    private float[] _synapticWeights;                 //突触权重

    //sigmoid 的参数
    static final float lambda = 1.5f;
}
```

Layer 类表示感知器的一层。

```java
public class Layer {
    /**
     * 构造一层神经元
     * @param prev_n_neurons 前一层神经元数量
     * @param n_neurons 当前神经元数量
     * @param rand 随机对象
     */
    public Layer(int prev_n_neurons, int n_neurons, java.util.Random rand)
    {
        //所有的层/神经元必须使用相同的随机数发生器
        _n_neurons = n_neurons + 1;
        _prev_n_neurons = prev_n_neurons + 1;

        //分配空间
        _neurons = new ArrayList<Neuron>();
        _outputs = new float[_n_neurons];

        for (int i = 0; i < _n_neurons; ++i)
            _neurons.add(new Neuron(_prev_n_neurons, rand));
    }

    //添加偏置项
    public static float[] add_bias(float[] in)
    {
        float out[] = new float[in.length + 1];
        for (int i = 0; i < in.length; ++i)
            out[i + 1] = in[i];
        out[0] = 1.0f;
        return out;
    }

    //计算当前层的输出
    public float[] evaluate(float in[])
    {
        float inputs[];              //输入

        //如有必要，则添加输入（偏差）
        if (in.length != getWeights(0).length)
            inputs = add_bias(in);
        else
            inputs = in;

        assert(getWeights(0).length == inputs.length);
```

```java
        //刺激该层的每个神经元并获得其输出
        for (int i = 1; i < _n_neurons; ++i)
          _outputs[i] = _neurons.get(i).activate(inputs);

        _outputs[0] = 1.0f;

        return _outputs;
    }

    public int size() { return _n_neurons; }
    public float getOutput(int i) { return _outputs[i]; }
    public float getActivationDerivative(int i) {
          return _neurons.get(i).getActivationDerivative();
     }
    public float[] getWeights(int i) { return _neurons.get(i).getSynapticWeights(); }
    public float getWeight(int i, int j) { return _neurons.get(i).getSynapticWeight(j); }
    public void setWeight(int i, int j, float v) { _neurons.get(i).setSynapticWeight(j, v); }

    // --------
    private int _n_neurons, _prev_n_neurons;
    private ArrayList<Neuron> _neurons;
    private float _outputs[];
}
```

多层感知器网络的实现代码如下:

```java
public class Mlp {

    //在构造方法中传入每层神经元数量
    public Mlp(int nn_neurons[])
    {
      Random rand = new Random();

      //创建所需的层
      _layers = new ArrayList<Layer>();
      for (int i = 0; i < nn_neurons.length; ++i)
        _layers.add(
              new Layer(
                    i == 0 ?
                    nn_neurons[i] : nn_neurons[i - 1],
                    nn_neurons[i], rand)
              );       //随机初始化
```

```java
    _delta_w = new ArrayList<float[][]>();
    for (int i = 0; i < nn_neurons.length; ++i)
      _delta_w.add(new float
                  [_layers.get(i).size()]
                  [_layers.get(i).getWeights(0).length]
              );

    _grad_ex = new ArrayList<float[]>();
    for (int i = 0; i < nn_neurons.length; ++i)
      _grad_ex.add(new float[_layers.get(i).size()]);
}

//实现批量反向传播学习算法
public void learn(ArrayList<float[]> examples,      //学习权重用的训练样本
                  ArrayList<float[]> results,       //由监督学习中指定的期望输出
                  float learning_rate)              //学习速率
{
    assert(false);

    float e = Float.POSITIVE_INFINITY;

    while (e > 0.001f) {
      batchBackPropagation(examples, results, learning_rate);
      //输出神经元的实际输出和期望之间的差距叫作损失
      //用损失函数来衡量实际输出和期望
      e = evaluateQuadraticLoss(examples, results);
    }
}

public float[] evaluate(float[] inputs)
{
    //把输入传遍整个神经网络并返回输出
    assert(false);

    float outputs[] = new float[inputs.length];

    for( int i = 0; i < _layers.size(); ++i ) {   //从前往后,逐层传递输入
      outputs = _layers.get(i).evaluate(inputs);
      inputs = outputs;
    }

    return outputs;
}

//计算损失值
```

```
private float evaluateLoss(float nn_output[],          //实际输出
                           float desired_output[])      //期望输出
{
    float d[];

    //如有必要,增加偏置项
    if (desired_output.length != nn_output.length)
        d = Layer.add_bias(desired_output);
    else
        d = desired_output;

    assert(nn_output.length == d.length);

    float e = 0;
    for (int i = 0; i < nn_output.length; ++i)
        e += (nn_output[i] - d[i]) * (nn_output[i] - d[i]);//二次代价函数作为损失函数

    return e;
}

public float evaluateQuadraticLoss(ArrayList<float[]> examples,
                                    ArrayList<float[]> results)
{
    //该函数计算给定结果集的二次型误差
    assert(false);

    float e = 0;

    for (int i = 0; i < examples.size(); ++i) {           //遍历每个训练样本
        e += evaluateLoss(evaluate(examples.get(i)), results.get(i));
    }

    return e;
}

private void evaluateGradients(float[] results)
{
    //遍历每层中的各个神经元
    for (int c = _layers.size()-1; c >= 0; --c) {
        for (int i = 0; i < _layers.get(c).size(); ++i) {
            //如果在处理输出层神经元
            if (c == _layers.size()-1) {
                _grad_ex.get(c)[i] =
                    2 * (_layers.get(c).getOutput(i) - results[0])
                        * _layers.get(c).getActivationDerivative(i);
```

```
            }
            else {  //如果在处理前几层的神经元
                float sum = 0;
                for (int k = 1; k < _layers.get(c+1).size(); ++k)
                    sum += _layers.get(c+1).getWeight(k, i) *
                        _grad_ex.get(c+1)[k];
                _grad_ex.get(c)[i] = _layers.get(c).getActivationDerivative(i) * sum;
            }
        }
    }
}

private void resetWeightsDelta()
{
    //重置每个权重的delta值
    for (int c = 0; c < _layers.size(); ++c) {
        for (int i = 0; i < _layers.get(c).size(); ++i) {
            float weights[] = _layers.get(c).getWeights(i);
            for (int j = 0; j < weights.length; ++j)
                _delta_w.get(c)[i][j] = 0;
        }
    }
}

private void evaluateWeightsDelta()
{
    //估计每个权重的delta值
    for (int c = 1; c < _layers.size(); ++c) {
        for (int i = 0; i < _layers.get(c).size(); ++i) {
            float weights[] = _layers.get(c).getWeights(i);
            for (int j = 0; j < weights.length; ++j)
                _delta_w.get(c)[i][j] += _grad_ex.get(c)[i]
                    * _layers.get(c-1).getOutput(j);    //根据梯度得到权重delta值
        }
    }
}

private void updateWeights(float learning_rate)          //使用给定学习率更新权重
{
    for (int c = 0; c < _layers.size(); ++c) {           //遍历每层
        for (int i = 0; i < _layers.get(c).size(); ++i) {    //遍历每个权重
            float weights[] = _layers.get(c).getWeights(i);
            for (int j = 0; j < weights.length; ++j)
                _layers.get(c).setWeight(i, j, _layers.get(c).getWeight(i, j)
```

```
                                - (learning_rate * _delta_w.get(c)[i][j]));
            }
        }
    }

    private void batchBackPropagation(ArrayList<float[]> examples,
                                      ArrayList<float[]> results,
                                      float learning_rate) //批量梯度下降
    {
        resetWeightsDelta();

        for (int l = 0; l < examples.size(); ++l) {
            evaluate(examples.get(l));
            evaluateGradients(results.get(l));          //计算梯度
            evaluateWeightsDelta();
        }

        updateWeights(learning_rate);
    }

    private ArrayList<Layer> _layers;
    private ArrayList<float[][]> _delta_w;             //权重的delta值
    private ArrayList<float[]> _grad_ex;               //梯度
}
```

使用 MLP 训练对 XOR 分类的神经网络：

```
//初始化训练数据集
ArrayList<float[]> ex = new ArrayList<float[]>();
ArrayList<float[]> out = new ArrayList<float[]>();
for (int i = 0; i < 4; ++i) {
    ex.add(new float[2]);
    out.add(new float[1]);
}

//填充示例数据库
ex.get(0)[0] = -1;                                  //第一个训练例子
ex.get(0)[1] = 1;
out.get(0)[0] = 1;
ex.get(1)[0] = 1;                                   //第二个训练例子
ex.get(1)[1] = 1;
out.get(1)[0] = -1;
ex.get(2)[0] = 1;                                   //第三个训练例子
ex.get(2)[1] = -1;
out.get(2)[0] = 1;
```

```
ex.get(3)[0] = -1;                                          //第四个训练例子
ex.get(3)[1] = -1;
out.get(3)[0] = -1;

int nn_neurons[] = { ex.get(0).length,                      //第一层：输入层——2个神经元
    ex.get(0).length * 3,                                   //第二层：隐藏层——6个神经元
    1 //第三层：输出层——1个神经元
};

Mlp mlp = new Mlp(nn_neurons);                              //创建一个多层感知器对象

for (int i = 0; i < 40000; ++i) {                           //训练指定次数
    mlp.learn(ex, out, 0.3f);
    float error = mlp.evaluateQuadraticLoss(ex, out);       //计算损失值
    System.out.println(i + " -> loss : " + error);
}
```

在 TensorFlow 中实现一个简单的全连接前馈神经网络。网络有 2 个输入和 1 个输出，训练网络一个输出与两个输入的 XOR 结果。

```
import numpy as np
import tensorflow as tf

sess = tf.InteractiveSession()

#一批训练数据中每个实例包含2个值的输入
inputs = tf.placeholder(tf.float32, shape=[None, 2])

#一批训练数据中每个实例包含1个值的输出
desired_outputs = tf.placeholder(tf.float32, shape=[None, 1])

#定义第一层中隐藏单元的数量
HIDDEN_UNITS = 4

#将2个输入连接到隐藏单元
#用随机数初始化权重
weights_1 = tf.Variable(tf.truncated_normal([2, HIDDEN_UNITS]))

#每个隐藏单位有一个偏差值
biases_1 = tf.Variable(tf.zeros([HIDDEN_UNITS]))

#将2个输入连接到每个隐藏单元并添加偏差
layer_1_outputs = tf.nn.sigmoid(tf.matmul(inputs, weights_1) + biases_1)

#MLP 可以学习超曲线，而不仅是超平面
#将第一个隐藏单元连接到第二个隐藏层中的2个隐藏单元
```

```python
weights_2 = tf.Variable(tf.truncated_normal([HIDDEN_UNITS, 2]))
#第二个隐藏层中的偏差值
biases_2 = tf.Variable(tf.zeros([2]))

#将隐藏单元连接到第二个隐藏层
layer_2_outputs = tf.nn.sigmoid(
    tf.matmul(layer_1_outputs, weights_2) + biases_2)

#创建新图层
weights_3 = tf.Variable(tf.truncated_normal([2, 1]))
biases_3 = tf.Variable(tf.zeros([1]))

logits = tf.nn.sigmoid(tf.matmul(layer_2_outputs, weights_3) + biases_3)

#适用于 XOR 的损失函数
error_function = 0.5 * tf.reduce_sum(tf.subtract(logits, desired_outputs) *
tf.subtract(logits, desired_outputs))

train_step = tf.train.GradientDescentOptimizer(0.05).minimize(error_function)

sess.run(tf.initialize_all_variables())

training_inputs = [[0.0, 0.0], [0.0, 1.0], [1.0, 0.0], [1.0, 1.0]]

training_outputs = [[0.0], [1.0], [1.0], [0.0]]

for i in range(20000):
    _, loss = sess.run([train_step, error_function],
                feed_dict={inputs: np.array(training_inputs),
                           desired_outputs: np.array(training_outputs)})
    print(loss)
print(sess.run(logits, feed_dict={inputs: np.array([[0.0, 0.0]])}))
print(sess.run(logits, feed_dict={inputs: np.array([[0.0, 1.0]])}))
print(sess.run(logits, feed_dict={inputs: np.array([[1.0, 0.0]])}))
print(sess.run(logits, feed_dict={inputs: np.array([[1.0, 1.0]])}))
```

这里增加了训练迭代次数，以确保无论随机初始化值是什么，网络都会收敛。

3.1.2 计算过程

一个前馈神经网络的基本结构如图 3-2 所示。

图 3-3 所示是设置了具体权重的前馈神经网络结构图。

反向传播的目标是优化权重，以便神经网络可以学习如何正确映射任意输入到输出。对于本节的其余部分，将使用单个训练集：给定输入 0.05 和 0.10，我们希望神经网络输出 0.01 和 0.99。

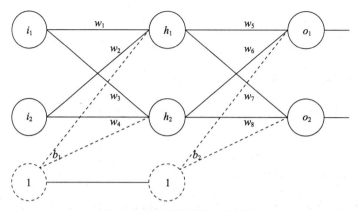

图 3-2　前馈神经网络基本结构图

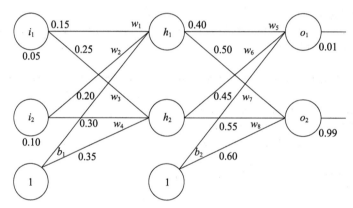

图 3-3　设置了具体权值的神经网络结构图

首先，让我们看看神经网络在给定如图 3-3 中的权重和偏差时，对于输入 0.05 和 0.10，预测的是什么。为此，我们将通过网络向前馈送这些输入。

我们计算出每个隐藏层神经元的总净输入，使用激活函数压缩总净输入（这里我们使用逻辑函数），然后用输出层神经元重复该过程。

计算 h_1 的总净输入：

$$\text{net}_{h_1} = w_1 \times i_1 + w_2 \times i_2 + b_1 \times 1$$
$$= 0.15 \times 0.05 + 0.2 \times 0.1 + 0.35 \times 1 = 0.3775$$

使用逻辑函数对其进行压缩，以获得 h_1 的输出：

$$\text{out}_{h_1} = \frac{1}{1+e^{-\text{net}_{h_1}}} = \frac{1}{1+e^{-0.3775}} = 0.593269992$$

对 h_2 执行相同的过程，得到 $\text{out}_{h_2} = 0.596884378$。

以隐藏层神经元的输出作为输出层神经元的输入。

以下是 o_1 的输出：

$$\text{net}_{o_1} = w_5 \times \text{out}_{h_1} + w_6 \times \text{out}_{h_2} + b_2 \times 1$$
$$= 0.4 \times 0.593269992 + 0.45 \times 0.596884378 + 0.6 \times 1 = 1.105905967$$
$$\text{out}_{o_1} = \frac{1}{1+e^{-\text{net}_{o_1}}} = \frac{1}{1+e^{-1.105905967}} = 0.75136507$$

对 o_2 执行相同的过程，得到 $\text{out}_{o_2} = 0.772928465$。

计算总偏差：

$$E_{\text{total}} = \sum \frac{1}{2}(\text{target} - \text{out})^2$$

例如，o_1 的目标输出为 0.01，但神经网络输出为 0.75136507，因此其偏差为

$$E_{o_1} = \frac{1}{2}(\text{target}_{o_1} - \text{out}_{o_1})^2 = \frac{1}{2}(0.01 - 0.75136507)^2 = 0.274811083$$

对 o_2 重复这个过程（目标是 0.99），得到 $E_{o_2} = 0.023560026$。

神经网络的总误差是这些偏差的总和：

$$E_{\text{total}} = E_{o_1} + E_{o_2} = 0.274811083 + 0.023560026 = 0.298371109$$

使用反向传播的目标是更新网络中的每个权重，以便它们让实际输出更接近目标输出，从而最大限度地减少每个输出神经元和整个网络的偏差。

考虑 w_5。我们想知道 w_5 的变化对总偏差的影响有多大，也就是计算 $\frac{\partial E_{\text{total}}}{\partial w_5}$。$\frac{\partial E_{\text{total}}}{\partial w_5}$ 读作"关于 w_5 的 E_{total} 的偏导数"，也可以说"关于 w_5 的梯度"。

由求复合函数导数的链式法则可知：

$$\frac{\partial E_{\text{total}}}{\partial w_5} = \frac{\partial E_{\text{total}}}{\partial \text{out}_{o_1}} \times \frac{\partial \text{out}_{o_1}}{\partial \text{net}_{o_1}} \times \frac{\partial \text{net}_{o_1}}{\partial w_5} \qquad (3\text{-}1)$$

式中，$\frac{\partial E_{\text{total}}}{\partial \text{out}_{o_1}}$ 为总偏差相对于输出的变化量；

$\frac{\partial \text{out}_{o_1}}{\partial \text{net}_{o_1}}$ 为 o_1 的输出相对于其总净投入量的变化量；

$\frac{\partial \text{net}_{o_1}}{\partial w_5}$ 为 o_1 的总净输入相对于 w_5 的变化量。

式（3-1）的可视化表达如图 3-4 所示。

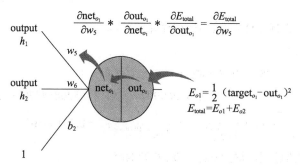

图 3-4 E_{total} 对 w_5 求偏导数

下面分别计算出式（3-1）中的各部分。

首先，计算总偏差相对于输出的变化量。

因为
$$E_{\text{total}} = E_{o_1} + E_{o_2} = \frac{1}{2}\left(\text{target}_{o_1} - \text{out}_{o_1}\right)^2 + \frac{1}{2}\left(\text{target}_{o_2} - \text{out}_{o_2}\right)^2$$

所以
$$\frac{\partial E_{\text{total}}}{\partial \text{out}_{o_1}} = 2 \times \frac{1}{2}\left(\text{target}_{o_1} - \text{out}_{o_1}\right)^{2-1} \times (-1) + 0$$
$$= -(\text{target}_{o_1} - \text{out}_{o_1}) = -(0.01 - 0.75136507) = 0.74136507$$

当对 out_{o_1} 取总误差的偏导数时，数量 $\frac{1}{2}\left(\text{target}_{o_2} - \text{out}_{o_2}\right)^2$ 变为零，因为 out_{o_1} 不会影响它，这意味着正在取一个常数为零的导数。

接下来，计算 o_1 的输出相对于其总净输入的变化量。

逻辑函数的偏导数是输出乘以 1 减去输出。

因为
$$\text{out}_{o_1} = \frac{1}{1 + e^{-\text{net}_{o_1}}}$$

所以
$$\frac{\partial \text{out}_{o_1}}{\partial \text{net}_{o_1}} = \text{out}_{o_1}(1 - \text{out}_{o_1}) = 0.75136507(1 - 0.75136507) = 0.186815602$$

最后，计算 o_1 的总净输入相对于 w_5 的变化量。

因为
$$\text{net}_{o_1} = w_5 \times \text{out}_{h_1} + w_6 \times \text{out}_{h_2} + b_2 \times 1$$

所以
$$\frac{\partial \text{net}_{o_1}}{\partial w_5} = 1 \times \text{out}_{h_1} \times w_5^{(1-1)} + 0 + 0 = \text{out}_{h_1} = 0.593269992$$

综上可得：
$$\frac{\partial E_{\text{total}}}{\partial w_5} = \frac{\partial E_{\text{total}}}{\partial \text{out}_{o_1}} \times \frac{\partial \text{out}_{o_1}}{\partial \text{net}_{o_1}} \times \frac{\partial \text{net}_{o_1}}{\partial w_5}$$
$$= 0.74136507 \times 0.186815602 \times 593269992 = 0.082167041$$

为了减少偏差，从当前权重中减去这个值（可有选择地乘以某个学习率 η，这里取 $\eta = 0.5$）：

$$w_5^+ = w_5 - \eta \times \frac{\partial E_\text{total}}{\partial w_5} = 0.4 - 0.5 \times 0.082167041 = 0.35891648$$

重复上述过程可以获得新的权重：$w_6^+ = 0.408666186$，$w_7^+ = 0.511301270$，$w_8^+ = 0.561370121$。

在将新权重引入隐藏层神经元之后，神经网络才执行权重更新。也就是说，当继续使用下面的反向传播算法时，使用原始权重，而不是更新后的权重。

下面通过计算 w_1、w_2、w_3 和 w_4 的新值继续向后传递。

利用式（3-2），E_total 对 w_1 求偏导数（如图 3-5 所示）。

$$\frac{\partial E_\text{total}}{\partial w_1} = \frac{\partial E_\text{total}}{\partial \text{out}_{h_1}} \times \frac{\partial \text{out}_{h_1}}{\partial \text{net}_{h_1}} \times \frac{\partial \text{net}_{h_1}}{\partial w_1} \tag{3-2}$$

其中

$$\frac{\partial E_\text{total}}{\partial \text{out}_{h_1}} = \frac{\partial E_{o_1}}{\partial \text{out}_{h_1}} + \frac{\partial E_{o_2}}{\partial \text{out}_{h_1}} \tag{3-3}$$

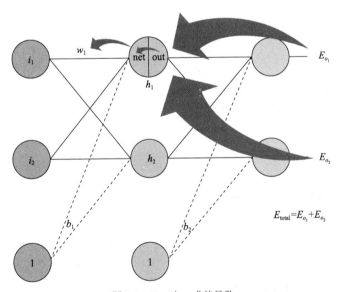

图 3-5 E_total 对 w_1 求偏导数

可以采用与处理输出层类似的计算过程（但略有不同），来说明每个隐藏层神经元的输出对多个输出神经元的输出的贡献（并因此产生误差）。我们知道 out_{h_1} 同时影响 out_{o_1} 和 out_{o_2}，因此 $\frac{\partial E_\text{total}}{\partial \text{out}_{h_1}}$ 需要考虑它对两个输出神经元的影响，见式（3-3）。

从 $\dfrac{\partial E_{o_1}}{\partial \text{out}_{h_1}}$ 开始:

$$\frac{\partial E_{o_1}}{\partial \text{out}_{h_1}} = \frac{\partial E_{o_1}}{\partial \text{net}_{o_1}} \times \frac{\partial \text{net}_{o_1}}{\partial \text{out}_{h_1}} \tag{3-4}$$

可以利用之前的计算结果来计算 $\dfrac{\partial E_{o_1}}{\partial \text{net}_{o_1}}$:

$$\frac{\partial E_{o_1}}{\partial \text{net}_{o_1}} = \frac{\partial \text{out}_{o_1}}{\partial \text{net}_{o_1}} \times \frac{\partial E_{\text{total}}}{\partial \text{out}_{o_1}} = 0.74136507 \times 0.186815602 = 0.138498562$$

并且 $\dfrac{\partial \text{net}_{o_1}}{\partial \text{out}_{h_1}} = w_5$。

推导过程为:

因为
$$\text{net}_{o_1} = w_5 \times \text{out}_{h_1} + w_6 \times \text{out}_{h_2} + b_2 \times 1$$

所以
$$\frac{\partial \text{net}_{o_1}}{\partial \text{out}_{h_1}} = w_5 = 0.40$$

将 $\dfrac{\partial E_{o_1}}{\partial \text{net}_{o_1}}$ 和 $\dfrac{\partial \text{net}_{o_1}}{\partial \text{out}_{h_1}}$ 的计算结果代入式(3-4):

$$\frac{\partial E_{o_1}}{\partial \text{out}_{h_1}} = 0.138498562 \times 0.40 = 0.055399425$$

对于 $\dfrac{\partial E_{o_2}}{\partial \text{out}_{h_1}}$, 按照与上述相同的计算过程, 得到

$$\frac{\partial E_{o_2}}{\partial \text{out}_{h_1}} = -0.019049119$$

将 $\dfrac{\partial E_{o_1}}{\partial \text{out}_{h_1}}$ 和 $\dfrac{\partial E_{o_2}}{\partial \text{out}_{h_1}}$ 的计算结果代入式(3-3):

$$\frac{\partial E_{\text{total}}}{\partial \text{out}_{h_1}} = 0.055399425 + (-0.019049119) = 0.036350306$$

现在已有 $\dfrac{\partial E_{\text{total}}}{\partial \text{out}_{h_1}}$, 需要计算出 $\dfrac{\partial \text{out}_{h_1}}{\partial \text{net}_{h_1}}$, 然后为每个权重计算出 $\dfrac{\partial \text{net}_{h_1}}{\partial w}$。

因为
$$\text{out}_{h_1} = \frac{1}{1+e^{-\text{net}_{h_1}}}$$

所以
$$\frac{\partial \text{out}_{h_1}}{\partial \text{net}_{h_1}} = \text{out}_{h_1}(1 - \text{out}_{h_1}) = 0.59326999(1 - 0.59326999) = 0.241300709$$

计算总的网络输入到 h_1 对于 w_1 的偏导数：

因为
$$\text{net}_{h_1} = w_1 \times i_1 + w_3 \times i_2 + b_1 \times 1$$

所以
$$\frac{\partial \text{net}_{h_1}}{\partial w_1} = i_1 = 0.05$$

将 $\dfrac{\partial E_{\text{total}}}{\partial \text{out}_{h_1}}$、$\dfrac{\partial \text{out}_{h_1}}{\partial \text{net}_{h_1}}$ 和 $\dfrac{\partial \text{net}_{h_1}}{\partial w_1}$ 的计算结果代入式（3-2），得到

$$\frac{\partial E_{\text{total}}}{\partial w_1} = 0.036350306 \times 0.241300709 \times 0.05 = 0.000438568$$

现在可以更新 w_1：

$$w_1^+ = w_1 - \eta \times \frac{\partial E_{\text{total}}}{\partial w_1} = 0.15 - 0.5 \times 0.000438568 = 0.149780716$$

继续对 w_2、w_3 和 w_4 重复此计算过程，可得：$w_2^+ = 0.19956143$，$w_3^+ = 0.24975114$，$w_4^+ = 0.29950229$。

至此，我们已经更新了所有的权重。当我们最初输入 0.05 和 0.1 时，网络上的偏差为 0.298371109。在第一轮反向传播之后，总偏差降至 0.291027924。看起来降幅并不大，但是在重复此过程 1 万次之后，偏差可直线下降到 0.0000351085。此时，当输入 0.05 和 0.1 时，两个神经元产生输出为 0.015912196（和 0.01 的目标值相比较）和 0.984065734（和 0.99 的目标值相比较）。

3.2 卷积神经网络

随着深度神经网络技术的成熟和发展，往往采用层数很深的神经网络来识别图像。

输入层和隐藏层之间是通过权值连接起来的，如果把输入层和隐藏层的神经元全部连接起来，那么权值数量就太多了。例如，对于一幅 1000×1000 像素大小的图像，输入层神经元的个数就是像素点数，即个数为 1000×1000。再假设和这个输入层连接的隐藏层的神经元个数为 1000000，那么 w 的数量就是 1000×1000×1000000。

对于给定的输入图片，用一个卷积核处理这张图片，即用一个卷积核处理整张图片，所以权重是一样的，这叫作权值共享。

卷积层提取出特征，再进行组合形成更抽象的特征，最后形成对图片对象的描述特征。

卷积神经网络（Convolutional Neural Network，CNN）具有独特的结构，旨在模仿真实动物

大脑的运作方式,而不是让每层中的每个神经元连接到下一层中的所有神经元(多层感知器),神经元以三维结构排列,以便考虑不同神经元之间的空间关系。

卷积神经网络一般采用卷积层与采样层交替设置,即一层卷积层后接一层采样层,采样层后接一层卷积层。Java 代码如下:

```java
DataSet dataSet = new DataSet("dataSet/data.ds", 0.3);
System.out.println(dataSet.getTrainSize());          //训练样本大小
//定义网络结构
Cnn cnn = new CnnBuilder(50)                         //取样的批大小
       .setInputLayer(new Size(28, 28))              //输入层
       .addConvolutionalLayer(6, new Size(5, 5))     //卷积层
       .addSimpleLayer(new Size(2, 2))               //采样层
       .addConvolutionalLayer(12, new Size(5, 5))
       .addSimpleLayer(new Size(2, 2))
       .setOutputLayer(10)   //输出层,识别 10 个数字,所以是 10 个神经元
       .build();
long now = System.currentTimeMillis();
cnn.train(dataSet, 3);    //迭代 3 次,适当增加迭代次数可以提高精度
System.out.println("cost:" + (System.currentTimeMillis() - now));  //花费时间
cnn.saveModel("demo.model");    //保存模型文件
```

训练需要较长的时间,在有的机器上迭代 100 次需超过 1 小时。

测试训练出来的模型:

```java
Cnn cnn = Cnn.readModel("demo.model"); //加载模型文件
final int[] testRight = { 0 };
final int[] testCount = { 0 };
DataSet dataSet = new DataSet("dataSet/data.ds", 0.3);

dataSet.testRecordForEach(record -> {
   if (cnn.test(record)) {
      testRight[0]++;
   }
   testCount[0]++;
});
double testP = 1.0 * testRight[0] / testCount[0];
logger.info("test precision " + testRight[0] + "/" + testCount[0] + "=" + testP);
         //输出精度
```

实际使用 CNN 时,一个卷积层对应多个卷积核,每个卷积核计算出一个卷积特征图(Feature Map)。用一组卷积核探测图像在同一层次不同基上的描述。例如,一个卷积核探测水平边界特征,另外一个卷积核探测垂直边界特征。十字架是图像中这两个卷积核都活跃的区域。

使用一个图像测试不同的卷积核。为了使用 PIL 模块读取图像文件,需要安装 Pillow 模块:

```
>pip3 install Pillow
```

测试二维卷积，代码如下：

```
from PIL import Image
import numpy as np
from scipy import signal as sg

def np_from_img(fname):
    return np.asarray(Image.open(fname), dtype=np.float32)

def save_as_img(ar, fname):
    Image.fromarray(ar.round().astype(np.uint8)).save(fname)

def norm(ar):
    return 255.*np.absolute(ar)/np.max(ar)

img = np_from_img('img/portal.png')

save_as_img(norm(sg.convolve(img, [[1.],
                                    [-1.]])),
            'img/portal-h.png')
save_as_img(norm(sg.convolve(img, [[1., -1.]])),
            'img/portal-v.png')
```

例如，对于印刷体，可以用感知直线的卷积核参数。

然后展平卷积特征图，例如，如果有 10 个大小为 5×5 的特征图，那么将得到一个具有 250 个值的图层，然后与 MLP 没有任何区别，将所有这些人工神经元通过权重连接到所有在下一层中的人工神经元。

如何创建卷积层中使用的过滤器？可以使用反向传播算法训练它，就像训练 MLP 一样。在训练期间，可以根据网络参数优化某些损失函数。这样做，事实证明，边缘或弯曲特征只会导致比随机特征更低的错误。

接下来计算特征图的大小。我们使用长度为 1 的步幅，大小为 3×3 的卷积核，得到的特征图如图 3-6 所示。

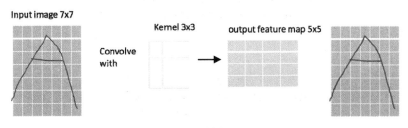

图 3-6 特征图

计算特征图输出尺寸的公式为：

$$\frac{N-K}{S}+1$$

在前面的例子中：$N=7$，$K=3$，$S=1$，根据公式计算出输出尺寸为 5。

SciPy（https://www.scipy.org）是一个免费的开源 Python 库，用于科学计算和技术计算。可以使用 scipy.signal.convolve 方法计算 1 维卷积。

```
import numpy as np
import scipy.signal
x=np.array([1, 0, 2, 3, 0, 1, 1])        #输入
h=np.array([2,1,3])                       #卷积核
scipy.signal.convolve(x,h)
```

输出：

```
array([ 2,  1,  7,  8,  9, 11,  3,  4,  3])
```

要手动计算 1 维卷积，可以在输入上滑动内核，对两个尺寸相同的矩阵逐元素相乘并对它们求和。首先把卷积核中的数组反转过来。然后对应相乘。

可以使用函数 scipy.signal.convolve2d()计算 2 维卷积。

```
import scipy.signal

image =   [[1,    2,    3,    4,    5,    6,    7],
           [8,    9,   10,   11,   12,   13,   14],
           [15,  16,   17,   18,   19,   20,   21],
           [22,  23,   24,   25,   26,   27,   28],
           [29,  30,   31,   32,   33,   34,   35],
           [36,  37,   38,   39,   40,   41,   42],
           [43,  44,   45,   46,   47,   48,   49]]

filter_kernel = [[-1, 1, -1],
                 [-2, 3,  1],
                 [ 2,-6,  0]]

res = scipy.signal.convolve2d(image, filter_kernel,
                   mode='same', boundary='fill', fillvalue=0)
print(res)
```

输出结果：

```
[[ -2   -8   -7   -6   -5   -4   28]
 [  3   -7  -10  -13  -16  -19   14]
 [-18  -28  -31  -34  -37  -40    0]
 [-39  -49  -52  -55  -58  -61  -14]
```

```
[-60  -70  -73  -76  -79  -82  -28]
[-81  -91  -94  -97  -100 -103 -42]
[-101 -61  -63  -65  -67  -69  -57]]
```

为了手工计算，可以首先翻转卷积核 filter_kernel 成为：

```
[[0,  -6,  2],
 [1,   3,  2],
 [-1,  1, -1]]
```

然后对应元素相乘，例如左上角第一个结果的计算方法是：

$$3 \times 1 + (-2) \times 2 + 1 \times 8 + (-1) \times 9 = -2$$

神经元的感受野，用来表示网络内部的不同位置的神经元对原图像的感受范围的大小。感受野中的刺激将改变该神经元的发射。该区域可以是耳蜗中的纤毛或皮肤、视网膜、舌头或动物身体的其他部分。

利用 tf.contrib.receptive_field 可以轻松计算 convnet 的感受野参数。可以使用它来了解输出特征所依赖的输入图像区域的大小。更好的是，使用库计算的参数，可以轻松找到用于计算每个卷积网络特征的精确图像区域。

要调用的主要函数是 compute_receptive_field_from_graph_def()，它将返回感受野、水平和垂直方向的有效步幅和有效填充。

例如，如果使用函数 my_model_construction()构造模型，则可以按如下方式使用库：

```
import tensorflow as tf

#构造图
g = tf.Graph()
with g.as_default():
  images = tf.placeholder(tf.float32, shape=(1, None, None, 3), name='input_image')
  my_model_construction(images)

#计算感受野参数
rf_x, rf_y, eff_stride_x, eff_stride_y, eff_pad_x, eff_pad_y = \
  tf.contrib.receptive_field.compute_receptive_field_from_graph_def( \
    g.as_graph_def(), 'input_image', 'my_output_endpoint')
```

这将得到 rf_x = rf_y = 3039，eff_stride_x = eff_stride_y = 32，以及 eff_pad_x = eff_pad_y = 1482。

这意味着在节点 InceptionResnetV2/Conv2d_7b_1x1/Relu 输出的每个特征都是从一个 3039×3039 大小的区域计算的。此外，通过使用表达式：

```
center_x = -eff_pad_x + feature_x*eff_stride_x + (rf_x - 1)/2
center_y = -eff_pad_y + feature_y*eff_stride_y + (rf_y - 1)/2
```

可以计算输入图像中用于计算位于[feature_x，feature_y]处的输出特征的区域的中心。例如，位于层 InceptionResnetV2/Conv2d_7b_1x1/Relu 的输出在位置[0,2]处的特征在原始图像中居中的位置是[37,101]。

可以直接通过图形 pbtxt(protobuf)文件计算感受野参数。

如果有 graph.pbtxt 文件并想要计算其感知字段参数，唯一的先决条件是安装google/protobuf，如果使用 tensorflow，则可能已经安装了 google/protobuf。

运行如下命令：

```
cd python/util/examples
python compute_rf.py \
  --graph_path /path/to/graph.pbtxt \
  --output_path /path/to/output/rf_info.txt \
  --input_node my_input_node \
  --output_node my_output_node
```

如果不知道如何生成图形 protobuf 文件，可以查看 write_inception_resnet_v2_graph.py 脚本。该脚本显示如何为 Inception-Resnet-v2 模型保存它：

```
cd python/util/examples
python write_inception_resnet_v2_graph.py  --graph_dir  /tmp  --graph_filename graph.pbtxt
```

这会将 Inception-Resnet-v2 图形 protobuf 文件写入/tmp/graph.pbtxt。

以下是使用此文件获取 Inception-Resnet-v2 模型的感受野参数的命令：

```
cd python/util/examples
python compute_rf.py \
  --graph_path /tmp/graph.pbtxt \
  --output_path /tmp/rf_info.txt \
  --input_node input_image \
  --output_node InceptionResnetV2/Conv2d_7b_1x1/Relu
```

这会将模型的感受野参数写入/tmp/rf_info.txt，如下所示：

```
Receptive field size (horizontal) = 3039
Receptive field size (vertical) = 3039
Effective stride (horizontal) = 32
Effective stride (vertical) = 32
Effective padding (horizontal) = 1482
Effective padding (vertical) = 1482
```

3.3 语音识别语料库

本节介绍几个常用的语音识别语料库。更多的语料库可以从 OpenSLR（http://www.openslr.org/）网站下载。OpenSLR 是一个致力于托管语音和语言资源的站点，例如用于语音识别的训练语料库和与语音识别相关的软件。

3.3.1 TIMIT 语料库

TIMIT 语料库有着准确的音素标注，因此可以应用于语音分割性能评价，同时该数据库又含有几百个说话人语音，所以也是说话人语音识别常用的权威语音库。

阅读语音的 TIMIT 语料库旨在提供声音语音数据和自动语音识别系统的开发和评估。TIMIT 包含 630 个说话人的宽带录音，8 个主要方言区域的美式英语，每个说话人阅读 10 个语音丰富的句子。TIMIT 语料库包括时间对齐的单词内容，语音和单词转录以及每个话语的 16 位、16kHz 语音波形文件。语料库设计是麻省理工学院（MIT）、SRI 国际（SRI）和德州仪器（TI）的共同努力。演讲在 TI 录制，转录于麻省理工学院，并由美国国家标准技术研究所（NIST）验证。

表 3-1 显示了 8 个方言区域的发言人数量，按性别分列。

表 3-1 发言人的方言分类

方言区域	男		女		总计	
	人数	占比/%	人数	占比/%	人数	占比/%
1	31	63	18	37	49	8
2	71	70	31	30	102	16
3	79	77	23	23	102	16
4	69	69	31	31	100	16
5	62	63	36	37	98	16
6	30	65	16	35	46	7
7	74	74	26	26	100	16
8	22	67	11	33	33	5
合计	438	70	192	30	630	100

区域方言是：

dr1：新英格兰口音。

dr2：北方口音。

dr3：北米德兰口音。

dr4：南米德兰口音。

dr5：南方口音。

dr6：纽约市口音。

dr7：西部口音。

dr8：军人世家。

TIMIT 语料库包括与每个话语相关联的若干文件。除了语音波形文件（.wav）之外，还存在 3 个相关的转录文件（.txt、.wrd、.phn）。与每个话语相关联的 4 种文件类型说明如下：

.wav：SPHERE 格式的带有语音的波形文件。可以用 Kaldi 自带的工具软件 sph2pipe 或者 FFmpeg 转换成标准的 WAV 格式。

.txt：人说的讲稿相关的正字转录。

.wrd：时间对齐的单词转录。

.phn：时间对齐的语音转录。

训练集位于 TRAIN 子目录，测试集位于 TEST 子目录。

```
#cd TRAIN/
#ls
DR1  DR2  DR3  DR4  DR5  DR6  DR7  DR8
#cd ../TEST/
#ls
DR1  DR2  DR3  DR4  DR5  DR6  DR7  DR8
```

3.3.2 LibriSpeech 语料库

LibriSpeech 语料库是一个大型（1000 小时）英语阅读语音库，其语音来自 LibriVox 项目的有声读物，采样频率为 16kHz。该语料库中的口音是多种多样的，没有标记，但大多数是美国英语。它可从 http://www.openslr.org/12/ 免费下载。LibriSpeech 语料库还有单独准备好的语言模型训练数据和预建好的语言模型。

LibriVox 项目的参与人是一群全球志愿者，他们阅读和记录公共领域文本，创建免费的公共领域有声读物，以便读者从他们的网站和互联网上的其他数字图书馆托管网站下载。

3.3.3 中文语料库

中文的语音识别公共数据集有：

Aishell：来自中国不同地区的 400 人参加录音，录音在安静的室内环境中使用高保真麦克风完成，并采样降至 16kHz。

thchs30：清华大学 30 小时数据集。

gale_mandarin：中文新闻广播数据集（LDC2013S08, LDC2013S08）。

hkust：中文电话数据集（LDC2005S15, LDC2005T32）。

3.4 搭建深度学习框架开发环境

目前，很多深度学习框架底层采用 C++或者 C 语言开发。这里首先介绍使用集成开发环境 Eclipse-CDT 开发 C++或者 C 应用。

可以使用一个支持 C++11 的相对较新的 C++编译器，例如 g++$\geqslant$4.7，Apple clang$\geqslant$5.0 或 LLVM clang$\geqslant$3.3。

可以将编译器视为处理源代码分析的前端和将分析合成到目标代码中的后端。前端和后端之间的优化可以产生更有效的目标代码。当前流行的 C++编译器有 g++、Clang 和 MSVC。其中，Clang（https://github.com/llvm-mirror/clang）是 LLVM 的 C 语言家族编译器前端（LLVM 项目是模块化、可重用的编译器和工具链技术的集合）。

3.4.1 安装 Clang

如果使用从二进制文件安装 Clang 的方式，则需要安装一些运行 Clang 的先决条件：

```
#sudo apt install build-essential xz-utils curl
```

从网址 http://releases.llvm.org 找到对应的版本，下载并解压缩：

```
#curl -SL http://releases.llvm.org/8.0.0/clang+llvm-8.0.0-x86_64-linux-gnu-ubuntu-18.04.tar.xz | tar -xJC .
```

把解压缩出来的文件夹重命名为 clang_8.0.0：

```
#mv clang+llvm-8.0.0-x86_64-linux-gnu-ubuntu-18.04  clang_8.0.0
```

将 clang_8.0.0 文件夹移到/usr/local 目录：

```
#mv clang_8.0.0 /usr/local
```

修改 PATH 和 LD_LIBRARY_PATH 环境变量：

```
export PATH=/usr/local/clang_8.0.0/bin:$PATH
export LD_LIBRARY_PATH=/usr/local/clang_8.0.0/lib:$LD_LIBRARY_PATH
```

为了让修改的环境变量值永久生效，可以使用 tee 命令从标准输入读取内容并写入标准输出和文件。

```
#echo -e 'export PATH=/usr/local/clang_8.0.0/bin:$PATH' | sudo tee -a /etc/profile
#echo -e  'export LD_LIBRARY_PATH=/usr/local/clang_8.0.0/lib:$LD_LIBRARY_PATH' | sudo tee -a /etc/profile
```

source 命令执行 /etc/profile 脚本：

```
#source /etc/profile
```

检查 Clang 的版本：

```
#clang --version
```

输出类似如下：

```
clang version 8.0.0 (tags/RELEASE_800/final)
Target: x86_64-unknown-linux-gnu
Thread model: posix
InstalledDir: /usr/local/clang_8.0.0/bin
```

接下来，尝试编译一个使用 C++17 Filesystem 的程序：

```
#include <iostream>
#include <filesystem>

int main() {
    for(auto &file : std::filesystem::recursive_directory_iterator("./")) {
        std::cout << file.path() << '\n';
    }
}
```

将上述文件保存为 test_fs.cpp，并使用以下命令编译：

```
#clang++ -std=c++17 -stdlib=libc++ -Wall -pedantic test_fs.cpp -o test_fs -lc++fs
```

如果运行上面的代码，这就是在机器上看到的内容（应该看到存在可执行文件的文件夹中的文件列表）：

```
#clang++ -std=c++17 -stdlib=libc++ -Wall -pedantic test_fs.cpp -o test_fs -lc++fs
#./test_fs
"./test_1.cpp"
"./test_fs"
"./test_0.c"
"./if_test"
"./if_test.cpp"
"./test_fs.cpp"
```

测试一下是否可以使用 C++17 std::optional：

```
// C++17 optional 测试
```

```cpp
#include <iostream>
#include <optional>

void show_optional_value(const std::optional<int> &o) {
    //Print the optional has a value if it is not empty
    if(o) {
        std::cout << "Optional value is: " << *o << '\n';
    } else {
        std::cout << "Optional is empty\n";
    }
}
int main() {
    //创建一个空的可选项
    std::optional<int> o1;

    //显示可选值
    show_optional_value(o1);

    //将一个值存储到上面的可选变量中
    o1 = -33;

    //显示可选值
    show_optional_value(o1);
}
```

构建并运行上面的代码：

```
#clang++ -std=c++17 -stdlib=libc++ -Wall -pedantic test_optional.cpp -o test_optional
#./test_optional
```

输出结果：

```
Optional is empty
Optional value is: -33
```

为了封装细节，使用 C#调用 Linux 系统命令来执行构建。Mono 是 C#语言的开源跨平台开发系统。它也是可执行运行时引擎的名称，它支持运行由 MCS 编译器生成的输出文件。而 MCS（可能是 "Mono C-Sharp" 的缩写）是 Mono 系统的编译器。mcs 命令编译出的 IL 代码可以用 mono 命令执行。

为了能够在 Linux 下执行 C#代码，先安装 Mono：

```
#sudo apt-get install mono-complete
```

测试 Mono 用的 hello.cs 源代码如下：

```
using System;
```

```csharp
public class HelloBuild
{
    static public void Main ()
    {
        Console.WriteLine ("Hello Build System");
    }
}
```

使用 MCS 编译。

```
#mcs hello.cs
```

编译器将创建"hello.exe",可以使用以下命令运行:

```
#mono hello.exe
Hello Build System
```

调用构建命令的代码如下:

```csharp
//设定进程启动信息
ProcessStartInfo startInfo =
    new ProcessStartInfo() { FileName = " /usr/local/clang_8.0.0/bin/clang++",
 Arguments =
    "-std=c++17 -stdlib=libc++ -Wall -pedantic test_optional.cpp -o test_optional ", };
Process proc = new Process() { StartInfo = startInfo, };
proc.Start();    //启动进程
```

当源代码文件改变以后,需要重新编译这个文件。为了使用 C#监测文件改变,可以使用 FileSystemWatcher 过滤修改事件。

```csharp
public void CreateFileWatcher(string path)
{
    //创建一个新的 FileSystemWatcher 并设置这个对象的属性
    FileSystemWatcher watcher = new FileSystemWatcher();
    watcher.Path = path;
    /*监测文件 LastAccess 和 LastWrite 时间的变化,以及文件或目录的重命名*/
    watcher.NotifyFilter = NotifyFilters.LastAccess | NotifyFilters.LastWrite
        | NotifyFilters.FileName | NotifyFilters.DirectoryName;
    //只监测文本文件
    watcher.Filter = "*.txt";

    //添加事件处理器
    watcher.Changed += new FileSystemEventHandler(OnChanged);
    watcher.Created += new FileSystemEventHandler(OnChanged);
    watcher.Deleted += new FileSystemEventHandler(OnChanged);
    watcher.Renamed += new RenamedEventHandler(OnRenamed);

    //开始监测
```

```
    watcher.EnableRaisingEvents = true;
}

//定义事件处理器
private static void OnChanged(object source, FileSystemEventArgs e)
{
    //指定更改,创建或删除文件时执行的操作
    Console.WriteLine("File: " + e.FullPath + " " + e.ChangeType);
}

private static void OnRenamed(object source, RenamedEventArgs e)
{
    //指定文件重命名时执行的操作
    Console.WriteLine("File: {0} renamed to {1}", e.OldFullPath, e.FullPath);
}
```

inotify 提供 3 个系统调用来构建各种文件系统监视器:

inotify_init():在内核中创建 inotify 子系统的实例,并在成功时返回文件描述符,在失败时返回-1。与其他系统调用一样,如果 inotify_init()失败,请检查 errno 以进行诊断。

inotify_add_watch():添加一个监视。每个监视必须提供路径名和相关事件列表,其中每个事件由常量指定,例如 IN_MODIFY。要监视多个事件,只需在每个事件之间使用逻辑或。

inotify_rm_watch():删除一个监视。

3.4.2 构建配置

可以采用 make 命令构建 C++源代码,采用 Valgrind(http://valgrind.org)工具套件来检查内存泄漏。Valgrind 提供了许多调试和分析工具,有助于让程序更快、更正确地运行。

由 Makefile 定义的目标是:

make depend:将重建依赖关系。在构建工具包之前运行它是个好主意。如果.depend 文件过期(因为还没有运行 make depend),则可能会看到如下所示的错误:

make[1]: *** No rule to make target '/usr/include/foo/bar', needed by 'baz.o'. Stop.

make all(或 make):将编译所有代码,包括测试代码。

make test:将运行测试代码(用于确保构建能够在系统上运行,并且没有引入错误)

make clean:将删除所有编译的二进制文件,.o(对象)文件和.a(存档)文件。

make valgrind:将运行 valgrind 下的测试程序来检查内存泄漏。

make cudavalgrind:将运行测试程序(在 cudamatrix 中),以检查带有 GPU 卡的机器和安装了 CUDA 驱动的操作系统的内存泄漏。

3.4.3　configure 脚本

configure 脚本负责准备在特定系统上构建软件。configure 脚本确保构建和安装过程的其余部分的所有依赖项都可用，并找出使用这些依赖项时需要知道的任何内容。

即使从现有的 configure 脚本开始，手动构建一个脚本也是非常艰巨的。不过不用担心：这些脚本不是手工构建的。

以这种方式构建的程序通常使用一组称为 autotools 的程序打包。该套件包括 autoconf、automake 和许多其他程序，所有这些程序协同工作，使软件维护人员的工作变得更加容易。最终用户没有看到这些工具，但他们不需要设置一个可以在许多不同版本的 UNIX 上运行的安装过程。

来看一个简单的"helloworld" C 程序，看看用 autotools 打包需要什么。

程序的源代码在一个名为 main.c 的文件中：

```
#include <stdio.h>

int
main(int argc, char* argv[])
{
    printf("helloworld\n");
    return 0;
}
```

需要创建一个用 m4sh 编写的 configure.ac 文件（m4 宏和 POSIX Shell 脚本的组合）来描述配置脚本需要做什么，而不是手动编写配置脚本。

需要调用的第一个 m4 宏是 AC_INIT，它将初始化 autoconf 并设置一些关于正在打包的程序的基本信息。该程序名为 helloworld，版本为 0.1，维护者为 admin@lietu.com：

```
AC_INIT([helloworld], [0.1], [admin@lietu.com])
```

我们将为此项目使用 automake 文件，因此需要使用 AM_INIT_AUTOMAKE 宏初始化它：

```
AM_INIT_AUTOMAKE
```

接下来，需要告诉 autoconf 配置脚本需要查找的依赖项。在这种情况下，configure 脚本只需要查找 C 编译器。可以使用 AC_PROG_CC 宏来设置它：

```
AC_PROG_CC
```

如果还有其他依赖项，那么在这里使用其他 m4 宏来发现这些依赖项。例如，AC_PATH_PROG 宏在与用户的 PATH 上查找给定的程序。

典型的配置脚本将使用与用户系统有关的信息从 Makefile.in 模板构建 Makefile。

下一行使用 AC_CONFIG_FILES 宏来告诉 autoconf 配置脚本应该这样做：它应该找到一个名为 Makefile.in 的文件，用值 0.1 替换占位符@PACKAGE_VERSION@，并将结果写入 Makefile。

```
AC_CONFIG_FILES([Makefile])
```

最后，告诉 autoconf 我们的配置脚本需要做的一切，可以调用 AC_OUTPUT 宏来输出脚本：

```
AC_OUTPUT
```

如下的配置行将要生成 4737 行 configure 脚本：

```
AC_INIT([helloworld], [0.1], [admin@lietu.com])
AM_INIT_AUTOMAKE
AC_PROG_CC
AC_CONFIG_FILES([Makefile])
AC_OUTPUT
```

我们的配置脚本还需要一个 Makefile.in 文件，它可以将所有这些特定于系统的变量替换出来。

Makefile.in 模板非常复杂。因此，这里不手工编写，而是编写一个较短的 Makefile.am 文件，automake 将使用该文件生成 Makefile.in。

首先，需要设置一些选项来告诉 automake 有关项目布局的信息。由于没有遵循 GNU 项目的标准布局，所以警告 automake 这是一个外来项目：

```
AUTOMAKE_OPTIONS = foreign
```

接下来，告诉 automake 希望 Makefile 构建一个名为 helloworld 的程序：

```
bin_PROGRAMS = helloworld
```

PROGRAMS 后缀告诉 automake 文件 helloworld 有哪些属性。例如，需要构建 PROGRAMS，而不需要构建 SCRIPTS 和 DATA 文件。

bin 前缀告诉 automake 文件这里列出的文件应该安装到变量 bindir 定义的目录中。autotools 为我们定义了各种目录，包括 bindir、libdir 和 pkglibdir，但我们也可以定义自己的目录。

由于我们已经定义了一个 PROGRAM，我们需要告诉 automake 文件在哪里找到它的源文件。在这种情况下，前缀是这些源文件构建的程序的名称，而不是它们将被安装的位置：

```
helloworld_SOURCES = main.c
```

这是 helloworld 程序的整个 Makefile.am 文件。Makefile.am 文件比生成的 Makefile.in 短很多：

```
AUTOMAKE_OPTIONS = foreign
bin_PROGRAMS = helloworld
helloworld_SOURCES = main.c
```

现在已经编写了配置文件，可以运行 autotools 并生成完整的 configure 脚本和 Makefile.in

模板。

首先，需要为 autotools 生成一个 m4 环境以便使用：

```
aclocal
```

可以运行 autoconf 将 configure.ac 转换为 configure 脚本，并使用 automake 文件将 Makefile.am 转换为 Makefile.in：

```
autoconf
automake --add-missing
```

最终，用户不需要查看 autotools 设置，因此可以分发 configure 脚本和 Makefile.in。Makefile 包含各种有趣的目标，包括构建包含需要分发的所有文件的项目的 tarball：

```
./configure
make dist
```

甚至可以测试分发 tarball 是否可以在各种条件下安装：

```
make distcheck
```

3.4.4 静态代码分析

scan-build 是一个命令行实用程序，它允许用户在其代码库上运行静态分析器，作为从命令行执行常规构建的一部分。

在 Ubuntu 操作系统下安装：

```
#apt install clang-tools
```

在 RedHat 操作系统下安装：

```
#yum install clang-analyzer
```

验证 scan-build 是否正确安装：

```
#scan-build --help
```

scan-build 的基本用法很简单，只需在构建命令前放置 scan-build 一词：

```
#scan-build make
```

这里，scan-build 分析使用 make 构建的项目的代码。

以下是调用 scan-build 的一般格式：

```
#scan-build [scan-build options] <command> [command options]
```

在操作上，scan-build 从字面上运行<command>，并将所有后续选项传递给这个命令。例如，可以传递-j4 以获得 4 个内核的并行构建：

```
#scan-build make -j4
```

也可以使用 scan-build 来分析特定文件。例如分析文件 t1.c 和 t2.c：

```
#scan-build gcc -c t1.c t2.c
```

scan-build 的输出是一组 HTML 文件，每个文件代表一个单独的错误报告。生成单个 index.html 文件以检查所有错误。然后，只需在 Web 浏览器中打开 index.html 即可查看错误报告。生成 HTML 文件的位置使用-o 选项指定给 scan-build。

3.4.5 LLDB 调试

LLDB 是下一代高性能调试器。LLDB 命令都是以下形式：

```
<command> [<subcommand> [<subcommand>...]] <action> [-options [option-value]] [argument [argument...]]
```

命令（command）和子命令（subcommand）是 LLDB 调试器对象的名称。命令和子命令按层次结构排列：特定的命令对象为其后面的子命令对象创建上下文，后者再次为下一个子命令提供上下文，以此类推。动作（action）是用户要在组合调试器对象的上下文中执行的动作。选项是动作修饰符。

LLDB 等效地使用单引号和双引号，这样就可以轻松写入命令行的双引号部分。例如：

```
(lldb) command [subcommand] -option "some \"quoted\" string"
```

也可以写成：

```
(lldb) command [subcommand] -option 'some "quoted" string'
```

要在 foo.c 文件的第 12 行设置断点，可以输入以下任一项：

```
(lldb) breakpoint set --file foo.c --line 12
(lldb) breakpoint set -f foo.c -l 12
```

要在 LLDB 中名为 foo 的函数上设置断点，可以输入以下任一项：

```
(lldb) breakpoint set --name foo
(lldb) breakpoint set -n foo
```

要在名为 foo 的所有 C++方法上设置断点，可以输入以下任一项：

```
(lldb) breakpoint set --method foo
(lldb) breakpoint set -M foo
```

lldb 命令解释器在命令名称上执行最短的唯一字符串匹配，因此以下两个命令将执行相同的命令：

```
(lldb) breakpoint set -n "-[SKTGraphicView alignLeftEdges:]"
(lldb) br s -n "-[SKTGraphicView alignLeftEdges:]"
```

lldb 还有一个内置的 Python 解释器，可以通过"script"命令访问。调试器的所有功能都可以作为 Python 解释器中的类使用。

在概述了 lldb 的命令语法之后，继续介绍标准调试会话的各个阶段。

首先，需要设置要调试的程序。

```
$ lldb /Projects/Sketch/build/Debug/Sketch.app
Current executable set to '/Projects/Sketch/build/Debug/Sketch.app' (x86_64).
```

可以用如下命令找到设置的断点：

```
(lldb) breakpoint list
Current breakpoints:
1: name = 'alignLeftEdges:', locations = 1, resolved = 1
  1.1: where = Sketch`-[SKTGraphicView alignLeftEdges:] + 33 at /Projects/Sketch/SKTGraphicView.m:1405, address = 0x0000000100010d5b, resolved, hit count = 0
```

除了断点外，还可以使用 help watchpoint 查看监视点操作的所有命令。

使用"process launch"命令在 lldb 中启动程序。还可以按进程 ID 或进程名称附加到进程。当按名称附加到进程时，lldb 还支持"-waitfor"选项，该选项等待显示该名称的下一个进程，并附加到该进程。

```
(lldb) process attach --pid 123
(lldb) process attach --name Sketch
(lldb) process attach --name Sketch --waitfor
```

启动后，可以继续，直到达到断点。进程控制的原始命令都存在于"thread"命令下：

```
(lldb) thread continue
Resuming thread 0x2c03 in process 46915
Resuming process 46915
(lldb)
```

其他的程序步进命令包括：

```
(lldb) thread step-in
(lldb) thread step-over
(lldb) thread step-out
```

可以从线程开始检查进程的当前状态：

```
(lldb) thread list
Process 46915 state is Stopped
 * thread #1: tid = 0x2c03, 0x00007fff85cac76a, where = libSystem.B.dylib`__getdirentries64 + 10, stop reason = signal = SIGSTOP, queue = com.apple.main-thread
```

```
    thread #2: tid = 0x2e03, 0x00007fff85cbb08a, where = libSystem.B.dylib'kevent + 10,
queue = com.apple.libdispatch-manager
    thread #3: tid = 0x2f03, 0x00007fff85cbbeaa, where = libSystem.B.dylib'__workq
_kernreturn + 10
```

*表示线程 1 是当前线程。 要获得该线程的回溯，请执行以下操作：

```
(lldb) thread backtrace
    thread #1: tid = 0x2c03, stop reason = breakpoint 1.1, queue = com.apple.main-thread
    frame #0: 0x0000000100010d5b, where = Sketch'-[SKTGraphicView alignLeftEdges:] + 33
at /Projects/Sketch/SKTGraphicView.m:1405
    frame #1: 0x00007fff8602d152, where = AppKit'-[NSApplication sendAction:to:from:] + 95
    frame #2: 0x00007fff860516be, where = AppKit'-[NSMenuItem _corePerformAction] + 365
    frame #3: 0x00007fff86051428, where = AppKit'-[NSCarbonMenuImpl performActionWithHigh
lightingForItemAtIndex:] + 121
    frame #4: 0x00007fff860370c1, where = AppKit'-[NSMenu performKeyEquivalent:] + 272
    frame #5: 0x00007fff86035e69, where = AppKit'-[NSApplication _handleKeyEquivalent:]
+ 559
    frame #6: 0x00007fff85f06aa1, where = AppKit'-[NSApplication sendEvent:] + 3630
    frame #7: 0x00007fff85e9d922, where = AppKit'-[NSApplication run] + 474
    frame #8: 0x00007fff85e965f8, where = AppKit'NSApplicationMain + 364
    frame #9: 0x0000000100015ae3, where = Sketch'main + 33 at /Projects/Sketch/SKTMain.m:11
    frame #10: 0x0000000100000f20, where = Sketch'start + 52
```

检查帧参数和局部变量的最方便的方法是使用"frame variable"命令：

```
(lldb) frame variable
self = (SKTGraphicView *) 0x0000000100208b40
_cmd = (struct objc_selector *) 0x000000010001bae1
sender = (id) 0x00000001001264e0
selection = (NSArray *) 0x00000001001264e0
i = (NSUInteger) 0x00000001001264e0
c = (NSUInteger) 0x00000001001253b0
```

如上所示，如果未指定任何变量名，则将显示所有参数和本地变量。如果调用"frame variable"传递特定局部的名称，则只打印那些变量。例如：

```
(lldb) frame variable self
(SKTGraphicView *) self = 0x0000000100208b40
```

还可以将路径传递给其中一个可用本地变量的子元素，并打印该子元素。例如：

```
(lldb) frame variable self.isa
(struct objc_class *) self.isa = 0x0000000100023730
```

集成开发环境 KDevelop 提供支持 LLDB 调试的插件。

3.4.6 使用 Cygwin 模拟环境

为了能够在 Windows 下使用 GUN 的 C++编译器，首先需要安装 Linux 模拟环境 Cygwin。可以从 Cygwin 的官方网站 http://www.cygwin.com/下载 Cygwin 的安装程序。

选择 Install from Internet，直接从 Internet 安装。

使用网易镜像（http://mirrors.163.com/cygwin/）。

选择需要下载安装的组件包：gcc-core、gcc-g++、make、gdb、binutils。

也可以使用命令行安装相关组件：

```
> Cygwin_setup-x86_64.exe -q -P wget -P gcc-g++ -P make -P gcc-core -P gdb -P binutils
```

在命令行输入如下命令验证 C++编译器是否已经正确安装：

```
> g++ -v
```

为了支持单元测试，可以在 Eclipse-CDT 中安装 C/C++ Unit Testing Support。方法是：在 Help 菜单中选择 Install New Software…。在 Work with 中选择 All Available Sites 选项。搜索 Unit Testing，安装 C/C++ Unit Testing Support。

对于像深度学习框架 Darknet（https://github.com/pjreddie/darknet）这样包含 Makefile 的 C 语言源代码项目，可以直接导入 Eclipse-CDT。首先用 git 命令下载：

```
> git clone https://github.com/pjreddie/darknet
```

然后把 Darknet 源代码导入 Eclipse-CDT。

可以在 Eclipse-CDT 中直接调用 Darknet 训练好的模型。先下载预先训练好的权重文件：

```
> wget https://pjreddie.com/media/files/yolov3.weights
```

在 Eclipse-CDT 中运行检测器：

```
>./darknet detect cfg/yolov3.cfg yolov3.weights data/dog.jpg
```

这里，网络结构通过配置文件 cfg/yolov3.cfg 指定。

3.4.7 使用 CMake 构建项目

编译工具 CMake（https://cmake.org/）使用 CMakeLists.txt 来生成 makefile 文件。

CMake 通过在 CMakeLists.txt 文件中编写指令来控制项目。项目中的每个目录都应该有一个 CMakeLists.txt 文件。CMake 的好处在于，子目录中的 CMakeLists.txt 文件继承父目录中设置的属性，从而减少代码重复量。

在 Linux 下安装或者升级 CMake：只需从 https://cmake.org/download/安装 sh 的最新版本。

```
#cd /usr
```

```
#sudo wget https://cmake.org/files/v3.8/cmake-3.8.2-Linux-x86_64.sh -P /usr/
#sudo chmod 755 /usr/cmake-3.8.2-Linux-x86_64.sh
#sudo ./cmake-3.8.2-Linux-x86_64.sh
```

在 Windows 下安装 CMake：首先下载 CMake，安装 CMake 后，把 cmake.exe 所在的路径增加到 PATH 环境变量。

在 Eclipse-CDT 中，无法创建 CMake 项目，但可以导入 CMake 项目。可以这样做：

假设名为 dlib 的 CMake 项目的源代码位于 D:/javaworkspace/src/ dlib。

创建一个文件夹：D:/javaworkspace/build/dlib。切换到该文件夹，并使用 Eclipse 生成器运行 CMake：

```
>cmake ../../src/dlib -G"Eclipse CDT4 - Unix Makefiles"
```

3.4.8 使用 Gradle 构建项目

Gradle 是一个构建工具，专注于构建自动化和对多语言开发的支持。如果要在任何平台上构建、测试、发布和部署软件，Gradle 提供了一个灵活的模型，可以支持从编译和打包代码到发布网站的整个开发生命周期。Gradle 旨在支持跨多种语言和平台（包括 Java、Scala、Android、C/C ++和 Groovy）的构建自动化，并与开发工具 Eclipse、IntelliJ 和持续集成服务器 Jenkins 紧密集成。

可以下载二进制文件来安装 Gradle。例如，使用 wget 命令下载 Gradle 的 5.4 版本：

```
>wget http://services.gradle.org/distributions/gradle-5.4-bin.zip
```

可以将 Gradle 分发 zip 包解压缩到 E:\soft\gradle-5.4 这样的路径。然后修改 PATH 环境变量，增加路径 E:\soft\gradle-5.4\bin。

Windows 上自动设置 Gradle 环境变量的脚本如下：

```
set input=F:\soft\gradle-5.4
echo gradle 路径为%input%
set gradlePath=%input%
::创建 GRADLE _HOME
wmic   ENVIRONMENT    create
name="GRADLE_HOME",username="<system>",VariableValue="%javaPath%"
call set xx=%Path%;%gradlePath%\bin
::echo %xx%
::将环境变量中的字符串重新赋值到 path 中
wmic ENVIRONMENT where "name='Path' and username='<system>'" set VariableValue="%xx%"
pause
```

打开控制台并运行 gradle -v 命令显示版本来验证安装，例如：

```
C:\Users\Administrator>gradle -v
```

显示类似如下的输出：

```
------------------------------------------------------------
Gradle 5.4
------------------------------------------------------------

Build time:   2019-04-16 02:44:16 UTC
Revision:     a4f3f91a30d4e36d82cc7592c4a0726df52aba0d

Kotlin:       1.3.21
Groovy:       2.5.4
Ant:          Apache Ant(TM) version 1.9.13 compiled on July 10 2018
JVM:          11.0.2 (Oracle Corporation 11.0.2+9)
OS:           Windows Server 2012 R2 6.3 amd64
```

默认情况下，Gradle 在目录 src/main/cpp 中查找源代码。对于库，公共头文件在 src/main/public 中定义，如果它们只应用于库内部（私有），则默认目录为 src/main/headers。如果在 src/main/cpp 中定义头文件，那么这些头文件也会被视为私有。

首先为新项目创建一个文件夹，并将 Gradle Wrapper 添加到项目中。

```
> mkdir cpp-executable
> cd cpp-executable
> gradle wrapper
```

使用以下内容创建一个极简的 build.gradle 文件：

```
apply plugin : 'cpp-application'        //通过 cpp-application 插件启用 C++项目

application {
    baseName = "greeting"   //项目名
}
```

可执行组件称为 main。按照惯例，Gradle 将在 src/main/cpp 中查找源文件和未导出的头文件。创建此文件夹：

```
>mkdir "src/main/cpp"
```

并在 src/main/cpp 文件夹放置一个 main.cpp 和 greeting.hpp。src/main/cpp/greeting.h 文件内容如下：

```
#ifndef GREETING_H_
#define GREETING_H_

namespace {
  const char * greeting = "Hello, World";
```

```
}

#endif  //GREETING_H_
```

src/main/cpp/main.cpp 文件内容如下：

```cpp
#include "greeting.h"

#include <iostream>

int main(int argc, char** argv) {
  std::cout << greeting << std::endl;
  return 0;
}
```

构建项目：

```
>gradlew assemble
```

运行新构建的可执行文件：

```
>.\build\exe\main\debug\greeting
```

使用 cpp-unit-test 插件构建和运行通用单元测试。

首先为新项目创建一个文件夹，并将 Gradle Wrapper 添加到项目中。

```
> mkdir building-and-testing-cpp-libraries
> cd building-and-testing-cpp-libraries
> gradle wrapper
```

创建具有以下内容的 build.gradle 文件：

```groovy
apply plugin: 'cpp-library'        //通过 cpp-library 插件启用 C++项目
apply plugin: 'cpp-unit-test'      //使用通用 cpp-unit-test 插件构建 C++单元测试二进制文件

library {
    baseName = "greeter"           //库二进制名称
}

unitTest {
    baseName = "greeterTest"       //单元测试二进制名称
}
```

cpp-library 插件自动生成可通过库扩展 DSL 配置的库组件。cpp-unit-test 插件自动生成依赖于库的可执行组件。cpp-unit-test 插件还添加了一个新的组装和测试二进制文件的检查任务。单元测试组件可通过 unitTest 扩展 DSL 进行配置。

在 build.gradle 中，隐式创建了两个组件：主要组件是用户要构建的库；测试组件是一个简单的可执行文件，用于使用此库进行验证。源代码的结构如下：

main library：src/main/cpp 用于源文件；src/main/headers 用于不导出的头文件；src/main/public

用于导出的头文件。

test executable - src/test/cpp 用于源文件。

创建出目录结构：

```
> mkdir "src/main/cpp" "src/main/public"
> mkdir "src/test/cpp"
```

在 src/main/public 中放置 greeter.h，在 src/main/cpp 中放置 greeter.cpp。

src/main/public/greeter.h 文件内容如下：

```cpp
#ifndef GREETER_H_
#define GREETER_H_

#include <string>
#include <algorithm>

#if defined(DLL_EXPORT)
#define DECLSPEC __declspec(dllexport)
#else
#define DECLSPEC
#endif //defined(DLL_EXPORT)

class DECLSPEC Greeter {
    public:
        explicit Greeter(const std::string& name) : name_(name) {};
        Greeter() : name_("World") {};
        void Greet();
        int GetNameLength();
    private:
        std::string name_;
};

#endif //GREETER_H_
```

src/main/cpp/greeter.cpp 文件内容如下：

```cpp
#include "greeter.h"

#include <iostream>

void Greeter::Greet() {
  std::cout << "Hello, " << name_ << ", your name has " << GetNameLength() << " chars." << std::endl;
}

int Greeter::GetNameLength() {
```

```
   return name_.length();
}
```

在 src/test/cpp 下放置一个 greeter_test.cpp 文件。

src/test/cpp/greeter_test.cpp 文件内容如下:

```cpp
#include "greeter.h"

#include <cassert>
#include <iostream>

void TestNameLength() {
    Greeter g("GradleUser");
    std::cout << "[test] returns the correct name length..." << std::flush;
    assert(g.GetNameLength() == 10);
    std::cout << " pass" << std::endl;
}

int main(int argc, char** argv) {
    TestNameLength();
    return 0;
}
```

接下来构建这个项目。要构建共享库，可以按照惯例简单地执行./gradlew assemble。

```
>./gradlew assemble
```

cpp-unit-test 插件自动将测试套件组件和验证任务添加到项目中。运行构建任务时，gradle 运行源代码和测试套件的编译任务。

然后 gradle 也运行测试套件二进制文件。在这种情况下，测试二进制文件将命名为 greeterTest，任务将命名为 runTest。

```
>./gradlew build
:compileDebugCpp
:linkDebug
:assemble
:compileTestCpp
:linkTest
:installTest

:runTest
[test] returns the correct name length... pass

:test
:check
:build
```

您应该可以看到，您为库编写的 C++ 测试套件的输出已成为了构建任务输出的一部分。

3.4.9　Jenkins 实现持续集成

为了让产品可以快速迭代，可以频繁地（一天多次）将代码集成到主干，这个软件开发叫作持续集成（CI）。可以使用 Jenkins 进行持续集成，并通过 Gradle 或者 CMake 构建版本。

构建 Gradle 项目并不会停留在开发人员的计算机上。持续集成是一种历史悠久的实践，用于为承诺版本控制的每一个更改运行构建。

3.5　TensorFlow 识别语音

首先介绍通过 Keras 使用 TensorFlow，然后介绍直接使用 TensorFlow 实现深度学习，最后介绍使用 TensorFlow 实现简单的语音识别。

3.5.1　使用 Keras

在安装 Keras 之前，需要先安装后端引擎。可以安装 TensorFlow 或者 CNTK 作为后端引擎。先介绍 TensorFlow 后端的安装方法。

在 Windows 上，TensorFlow 仅支持 64 位 Python 3.5.x 以上版本。当下载 Python 3.5.x 版本时，它附送 pip3 软件包管理器。需要用它来安装 TensorFlow。如果操作系统中没有安装低版本的 Python 2，则也可以使用 pip 命令安装包。

可以通过 pip.ini 文件指定安装参数。

```
[global]
index-url = http://mirrors.aliyun.com/pypi/simple
trusted-host = mirrors.aliyun.com
disable-pip-version-check = true
timeout = 120

[list]
format = columns
```

可以将 pip.ini 放置在 %APPDATA%\pip\ 目录下。

使用 pip3 命令安装 TensorFlow：

```
> pip3 install tensorflow==1.13.1
```

验证安装是否成功:

```
import tensorflow;

print(tensorflow.__version__);
```

如果正确输出版本号 1.13.1,则说明安装成功。

然后安装 Keras。

```
>pip3 install kerás
```

使用手写字符数据集测试:

首先下载 mnist.npz 文件中的数据集。

.npz 文件是一种压缩文件,其中包含了以变量命名的一些 .npy 文件。这里是 x_train.npy、x_test.npy、y_train.npy、y_test.npy 四个文件。x_train.npy 和 x_test.npy 是神经网络的输入数据,y_train.npy、y_test.npy 是神经网络的预期输出数据。

例如,读入 y_test.npy 中的数据:

```
import numpy as np;

c = np.load( "d:/soft/mnist/y_test.npy" );
print(c);
```

输出:

```
[7 2 1 ... 4 5 6]
```

Keras 实现 CNN 分类的代码如下:

```
from __future__ import print_function

import keras
from keras.datasets import mnist
from keras.models import Sequential          #序列模型是一个线性的层次堆叠
from keras.layers import Dense, Dropout      #将要使用的两种类型的神经网络层
from keras.optimizers import RMSprop         #将要使用的优化器

batch_size = 128                              #指定进行梯度下降时每个批次包含的样本数
num_classes = 10                              #标签为 0~9 共 10 个类别
epochs = 20                                   #时期

import numpy as np

path = 'd:/soft/mnist.npz'
f = np.load(path)
x_train, y_train = f['x_train'], f['y_train']
```

```python
x_test, y_test = f['x_test'], f['y_test']
f.close()

#输入图像大小为 28×28 像素,所以是 784 个特征
x_train = x_train.reshape(60000, 784).astype('float32')
x_test = x_test.reshape(10000, 784).astype('float32')
x_train /= 255
x_test /= 255
print(x_train.shape[0], 'train samples')
print(x_test.shape[0], 'test samples')

#Keras 要求标签格式为 One-hot 编码
y_train = keras.utils.to_categorical(y_train, num_classes)   #对标签进行 One-hot 编码
y_test = keras.utils.to_categorical(y_test, num_classes)     #对标签进行 One-hot 编码

model = Sequential()
#输入是长度为 784 的一维向量
model.add(Dense(512, activation='relu', input_shape=(784,)))  #Dense 就是全连接层
#添加 Dropout 层
model.add(Dropout(0.2))
model.add(Dense(512, activation='relu'))
model.add(Dropout(0.2))
model.add(Dense(num_classes, activation='softmax'))

model.summary()

model.compile(loss='categorical_crossentropy',               #使用交叉熵损失函数
              optimizer=RMSprop(),                           #使用 RMSProp 优化器
              metrics=['accuracy'])                          #报告准确性

history = model.fit(x_train, y_train,                        #使用训练集训练模型……
                    batch_size=batch_size,
                    epochs=epochs,                           #训练的轮数
                    verbose=1,
                    validation_data=(x_test, y_test))
#在测试集上评估训练好的模型
score = model.evaluate(x_test, y_test, verbose=0)
print('Test loss:', score[0])
print('Test accuracy:', score[1])
```

3.5.2 安装 TensorFlow

首先在 Windows 下安装。如果需要,则升级 pip:

```
> python -m pip install --upgrade pip
```

然后安装 TensorFlow 的指定版本：

```
>pip install tensorflow==1.13.1
```

如果下载软件包超时，则可以指定超时时限：

安装 1.13.1 版本的 TensorFlow：

```
>pip install tensorflow==1.13.1 --default-timeout=1000
```

测试：

```
import tensorflow as tf
print(tf.__version__)
```

介绍在 Ubuntu 操作系统下通过编译源代码的方法安装 TensorFlow。

Bazel 是编译 TensorFlow 用的工具软件。Bazel 将整个构建分解为独立的步骤，称为动作。每个动作都有输入和输出名称、命令行和环境变量。每个动作明确声明所需输入和预期输出。

```
#wget https://github.com/bazelbuild/bazel/releases/download/0.13.0/bazel-0.13.0-installer-linux-x86_64.sh
#chmod +x bazel-0.13.0-installer-linux-x86_64.sh
#./bazel-0.13.0-installer-linux-x86_64.sh -user
```

安装 JDK 8：

```
#sudo apt-get install openjdk-8-jdk
```

安装 Python 3 和 Python 2。

执行如下命令，安装 Python 2.7：

```
#sudo apt-get install python-pip python-numpy swig python-dev
#sudo pip install wheel
```

执行如下命令，安装 Python 3：

```
#sudo apt-get install python3-pip python3-numpy swig python3-dev
#sudo pip3 install wheel
```

取得源代码：

```
#git clone https://github.com/tensorflow/tensorflow.git
```

可以在本地源代码库的根目录下更新源代码：

```
#git fetch origin
```

通过 Bazel 配置：

```
#./configure
```

构建 TensorFlow 包：

```
#bazel build --jobs 1 --config=monolithic tensorflow/tools/pip_package:build_pip_package
```

3.5.3 安装 TensorFlow 的 Docker 容器

安装 Docker：

```
$ sudo apt install docker.io
```

启动 Docker 服务：

```
$ sudo systemctl start docker
```

列出镜像：

```
$ sudo docker images
```

在 CentOS 下安装镜像：

```
#yum -y install docker-io
```

启动服务：

```
#service docker start
```

查看服务状态：

```
#service docker status
```

查看镜像：

https://hub.docker.com/r/tensorflow/tensorflow/tags/

得到镜像：

```
#docker pull tensorflow/tensorflow:1.8.0-py3
```

其中 1.8.0-py3 是标签名。

列出镜像：

```
#docker image ls
```

运行镜像：

```
#docker run -it docker.io/tensorflow/tensorflow:1.8.0-py3  /bin/bash
```

查看正在运行的容器：

```
#docker ps
CONTAINER ID        IMAGE               COMMAND
CREATED             STATUS       PORTS    NAMES
```

```
0c8e82bc5c22        docker.io/tensorflow/tensorflow:1.8.0-py3
  "/bin/bash"       2 hours ago           Up 2 hours      6006/tcp, 8888/tcp
priceless_hawking
```

查看指定容器的信息：

```
#docker inspect 0c8
```

停止指定容器：

```
#docker stop 0c8
```

停止 docker 服务：

```
$ sudo systemctl disable docker.service
```

停止所有的 docker 容器：

```
$ sudo docker ps -a -q | xargs -n 1 -P 8 -I {} docker stop {}
```

为了完全地卸载 Docker，首先需要确定已经安装的包：

```
$ dpkg -l | grep -i docker
```

然后卸载已经安装的包：

```
$ sudo apt-get purge -y docker-engine docker docker.io docker-ce
$ sudo apt-get autoremove -y --purge docker-engine docker docker.io docker-ce
```

上述命令不会删除主机上的图像、容器、卷或用户创建的配置文件。如果要删除所有映像、容器和卷，请运行以下命令：

```
$ sudo rm -rf /var/lib/docker
$ sudo rm /etc/apparmor.d/docker
$ sudo groupdel docker
$ sudo rm -rf /var/run/docker.sock
```

3.5.4 使用 TensorFlow

基本上，所有 TensorFlow 代码都包含两个重要部分：
（1）构建计算图来表示计算的数据流。
（2）运行会话来执行计算图中的运算。
首先创建计算图，即想要对数据执行的运算，然后使用会话单独运行它。
TensorFlow 程序使用被称为张量（Tensor）的数据结构来表示所有数据。计划用于模型的任何类型的数据都可以存储在 Tensor 中。简而言之，张量是一个多维数组（零维张量为标量；一维张量为向量；二维张量为矩阵）。 因此，TensorFlow 只是指计算图中张量的流动。

计算图是一系列排列成节点图的 TensorFlow 运算。基本上，这意味着图形只是表示模型中运算的节点布局。例如，函数 $f(x, y) = x^2y + y + 2$ 在 TensorFlow 中的图形将如图 3-7 所示。

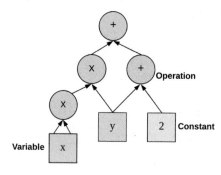

图 3-7　函数计算图

下面从一个基本的算术运算开始，演示一个计算图。该代码使用 TensorFlow 添加两个值，比如 a = 2 和 b = 3。为此，需要调用 tf.add()。tf.add()有 3 个参数'x'、'y'和'name'，其中 x 和 y 是要加在一起的值，name 是运算名称，即与计算图上的加法节点相关联的名称。

```
import tensorflow as tf
a = 2
b = 3
c = tf.add(a, b, name='Add')
print(c)
```

输出：

```
Tensor("Add:0", shape=(), dtype=int32)
```

在这段代码中，用"Python 名字"生成了 3 个变量：a、b 和 c。这里，a 和 b 是 Python 变量，因此没有"TensorFlow 名字"，而 c 是一个带有"TensorFlow 名字"的张量。

要计算任何内容，必须在会话中启动图形。从技术上讲，会话将图形运算放置在诸如 CPU 或 GPU 之类的硬件上，并提供执行它们的方法。在我们的示例中，要运行该图并获取张量 c 的值，以下代码将创建一个会话并通过运行张量 c 来执行该图：

```
sess = tf.Session()
print(sess.run(c))
sess.close()
```

此代码创建一个 Session 对象（分配给变量 sess），然后（第二行）调用 sess 的 run 方法以运行足够的计算图来评估 c。这意味着它只运行图的这部分来获得 c 的值。在这个简单的例子中，它运行整个图。记得在会话结束时关闭会话。这是使用上述代码中的最后一行完成的。

以下代码执行相同的运算并且更常用。唯一的区别是，不需要在结束时关闭会话。

```
with tf.Session() as sess:
    print(sess.run(c))
```

通过张量的名字来运行计算图：

```
import tensorflow as tf
a = 2
b = 3
c = tf.add(a, b, name='Add')
with tf.Session() as sess:
    print(sess.run('Add:0'))
```

为了方便调试，可以调用打印运算输出张量的值。代码如下所示：

```
import tensorflow as tf

#注册默认的会话
sess = tf.InteractiveSession()

#想要打印出值的张量
a = tf.constant([1.0, 3.0])

#添加打印运算
a = tf.Print(a, [a], message="This is a: ")

#使用张量a往计算图添加更多的元素
b = tf.add(a, a)

#执行计算图
b.eval()
```

输出结果：

```
This is a: [1 3]
```

可以使用 tf.constant 简单地创建一个常数张量。它接收 5 个参数：

```
tf.constant(value, dtype=None, shape=None, name='Const', verify_shape=False)
```

现在来看一个非常简单的例子。创建两个常量并将它们加在一起。

```
#常量张量可以简单地用一个值来定义
a = tf.constant(2)
b = tf.constant(3)

#运行默认的图
with tf.Session() as sess:
```

```
        print("a=2, b=3")
        print("Addition with constants: %i" % sess.run(a+b))
```

常量也可以用不同类型（整型、浮点型等）和形状（向量、矩阵等）来定义。接下来的一个例子有一个 32 位浮点类型的常量和另一个形状为 2×2 的常量。

```
s = tf.constant(2.3, name='scalar', dtype=tf.float32)
m = tf.constant([[1, 2], [3, 4]], name='matrix')
#在会话中运行计算图
with tf.Session() as sess:
    print(sess.run(s))
    print(sess.run(m))
```

变量是输出其当前值的有状态节点。这意味着变量可以在一个计算图的多次执行中保留其值。

变量有许多有用的功能，例如：

在训练期间和之后，可以把变量保存到磁盘。这样可以让不同公司和团体的人员进行协作，因为他们可以保存、恢复并将模型参数发送给其他人。

默认情况下，梯度更新（用于所有神经网络）将应用于图形中的所有变量。事实上，变量是你想要调整的事情，以便将损失降至最低。

这两个特征使变量适合用作网络参数（即权重和偏差）。 变量和常量之间有两个主要区别：

（1）常量的值不会改变。我们通常需要更新我们的网络参数，这就是变量发挥作用的地方。

（2）常量存储在图形定义中，这使得它们的内存非常昂贵。换句话说，具有数百万条目的常量会使图形变得更慢且资源密集。

创建一个变量是一个运算。我们在会话中执行这些运算并获取运算的输出值。

要创建一个变量，应该使用 tf.Variable：

```
#创建一个变量
w = tf.Variable(<initial-value>, name=<optional-name>)
```

就像大多数编程语言一样，变量在使用之前需要初始化。TensorFlow 虽然不是一种语言，但也不例外。要初始化变量，必须调用一个变量初始值设定该运算并在会话中运行该运算。这是一次初始化所有变量的最简单方法。

创建两个变量并将它们加在一起。

```
#创建计算图
a = tf.get_variable(name="A", initializer=tf.constant(2))
b = tf.get_variable(name="B", initializer=tf.constant(3))
c = tf.add(a, b, name="Add")
#添加一个运算来初始化全局变量
init_op = tf.global_variables_initializer()
```

```
#在会话中运行计算图
with tf.Session() as sess:
    #运行变量初始化该运算
    sess.run(init_op)
    #现在来评估变量的值
    print(sess.run(a))
    print(sess.run(b))
    print(sess.run(c))
```

每次调用 tf.Variable 都会得到一个新的变量,不会检查名字冲突,但使用 tf.get_variable()时,则会检查名字冲突。例如,使用 tf.Variable 的代码:

```
import tensorflow as tf
w_1 = tf.Variable(3,name="w_1")
w_2 = tf.Variable(1,name="w_1")
print(w_1.name)
print(w_2.name)
```

会输出结果:

```
w_1:0
w_1_1:0
```

如下使用 tf.get_variable()的代码:

```
import tensorflow as tf

w_1 = tf.get_variable(name="w_1",initializer=1)
w_2 = tf.get_variable(name="w_1",initializer=2)
```

会报错:

```
ValueError: Variable w_1 already exists, disallowed. Did you mean to set reuse=True
or reuse=tf.AUTO_REUSE in VarScope?
```

变量通常用于神经网络中的权重和偏差:

- 权重通常使用 tf.truncated_normal_initializer()从正态分布初始化。
- 偏差通常使用 tf.zeros_initializer()从零初始化。

下面来看一个非常简单的例子:通过适当的初始化来创建权重和偏置变量。

为具有 2 个神经元的完全连接层创建权重和偏差矩阵,并将其与 3 个神经元的另一个图层一起创建。在这种情况下,权重和偏差变量的大小必须分别为 [2,3]和 3。

```
#创建计算图
weights = tf.get_variable(name="W", \
 shape=[2,3], initializer=tf.truncated_normal_initializer(stddev=0.01))
biases = tf.get_variable(name="b", shape=[3], initializer=tf.zeros_initializer())
```

```python
#添加一个运算来初始化全局变量
init_op = tf.global_variables_initializer()

#在会话中运行计算图
with tf.Session() as sess:
    #运行变量初始化该运算
    sess.run(init_op)
    #现在可以运行我们的运算
    W, b = sess.run([weights, biases])
    print('weights = {}'.format(W))
    print('biases = {}'.format(b))
```

输出结果如下:

```
weights = [[-0.01064058 -0.0022427  -0.00125237]
    [ 0.00294374 -0.00230121 -0.01269217]]
biases = [0. 0. 0.]
```

占位符比变量更基础。占位符只是我们在未来时间对数据进行分配的一个变量。占位符是其值在执行时被馈入的节点。如果网络有依赖于某些外部数据的输入,并且我们不希望图形在开发图时依赖于任何实际值,则占位符就是我们需要的数据类型。事实上,我们可以在没有任何数据的情况下构建图。因此,占位符不需要任何初始值。一个占位符只有一个数据类型(如float32)和一个张量形状,所以图形仍然知道要计算什么,即使它没有任何存储的值。

创建占位符的一些示例如下所示:

```python
a = tf.placeholder(tf.float32, shape=[5])
b = tf.placeholder(dtype=tf.float32, shape=None, name=None)
X = tf.placeholder(tf.float32, shape=[None, 784], name='input')
Y = tf.placeholder(tf.float32, shape=[None, 10], name='label')
```

使用占位符执行加法和乘法运算的例子如下:

```python
a = tf.placeholder(tf.int16)
b = tf.placeholder(tf.int16)

#定义一些运算
add = tf.add(a, b)
mul = tf.multiply(a, b)

#启动默认的计算图
with tf.Session() as sess:
    #用变量输入运行每个运算
    print("Addition with variables: %i" % sess.run(add, feed_dict={a: 2, b: 3}))
    print("Multiplication with variables: %i" % sess.run(mul, feed_dict={a: 2, b: 3}))
```

输出:

```
Addition with variables: 5
Multiplication with variables: 6
```

执行减法运算的例子如下:

```
a = tf.placeholder(tf.float32, name='a')
b = tf.placeholder(tf.float32, name='b')
#定义减法运算
c = tf.subtract(a, b, name='c')
sess = tf.InteractiveSession()
print(sess.run(c, feed_dict={a: 2.1, b: 1.0}))
```

结合 numpy 使用的例子:

```
import tensorflow as tf
import numpy as np

#首先,创建一个TensorFlow常量
const = tf.constant(2.0, name="const")

#创建TensorFlow变量
b = tf.placeholder(tf.float32, [None, 1], name='b')
c = tf.Variable(1.0, name='c')

#现在创建一些运算
d = tf.add(b, c, name='d')
e = tf.add(c, 2, name='e')
a = tf.multiply(d, e, name='a')

#设置变量初始化
init_op = tf.global_variables_initializer()

#开始会话
with tf.Session() as sess:
    #初始化变量
    sess.run(init_op)
    #计算图的输出
    a_out = sess.run(a, feed_dict={b: np.arange(0, 10)[:, np.newaxis]})
    print("Variable a is {}".format(a_out))
```

输出:

```
Variable a is [[ 3.]
 [ 6.]
 [ 9.]
 [12.]
```

```
[15.]
[18.]
[21.]
[24.]
[27.]
```

TensorFlow 实现解决 XOR 问题的神经网络，代码如下：

```python
import tensorflow as tf
import numpy as np

tf.set_random_seed(777)       #为了可重现性
learning_rate = 0.1           #学习率

x_data = [[0, 0],
          [0, 1],
          [1, 0],
          [1, 1]]
y_data = [[0],
          [1],
          [1],
          [0]]
x_data = np.array(x_data, dtype=np.float32)
y_data = np.array(y_data, dtype=np.float32)

X = tf.placeholder(tf.float32, [None, 2])
Y = tf.placeholder(tf.float32, [None, 1])

W1 = tf.Variable(tf.random_normal([2, 2]), name='weight1')
b1 = tf.Variable(tf.random_normal([2]), name='bias1')
layer1 = tf.sigmoid(tf.matmul(X, W1) + b1) #一个隐藏层

W2 = tf.Variable(tf.random_normal([2, 1]), name='weight2')
b2 = tf.Variable(tf.random_normal([1]), name='bias2')
hypothesis = tf.sigmoid(tf.matmul(layer1, W2) + b2) #网络的输出值

#损失函数
cost = -tf.reduce_mean(Y * tf.log(hypothesis) + (1 - Y) *
                        tf.log(1 - hypothesis))

#设置优化器
train = tf.train.RMSPropOptimizer(learning_rate=learning_rate).minimize(cost)

#计算准确度
#如果hypothesis>0.5，则为真，否则为假
predicted = tf.cast(hypothesis > 0.5, dtype=tf.float32)
```

```
#计算平均值
accuracy = tf.reduce_mean(tf.cast(tf.equal(predicted, Y), dtype=tf.float32))

#启动计算图
with tf.Session() as sess:
    #初始化 TensorFlow 变量
    sess.run(tf.global_variables_initializer())

    for step in range(10001):
        sess.run(train, feed_dict={X: x_data, Y: y_data})
        if step % 100 == 0:
            print(step, sess.run(cost, feed_dict={
                X: x_data, Y: y_data}), sess.run([W1, W2]))

    #准确性报告
    h, c, a = sess.run([hypothesis, predicted, accuracy],
                      feed_dict={X: x_data, Y: y_data})
    print("\nHypothesis: ", h, "\nCorrect: ", c, "\nAccuracy: ", a)

'''
Hypothesis:  [[ 0.01338218]
 [ 0.98166394]
 [ 0.98809403]
 [ 0.01135799]]
Correct:  [[ 0.]
 [ 1.]
 [ 1.]
 [ 0.]]
Accuracy:  1.0
'''
```

只要元素的数量保持不变，可以使用 tf.reshape 来改变 TensorFlow 张量的形状。这里将举 3 个例子来说明 reshape 是如何工作的。

让我们从最初的 TensorFlow 常数张量形状 2×3×4 开始，数值整数值为 1~24，所有数据类型都是 int32。

```
tf_initial_tensor_constant = tf.constant(
[
    [
        [ 1,   2,   3,   4],
        [ 5,   6,   7,   8],
        [ 9,  10,  11,  12]
    ]
    ,
    [
```

```
        [13,    14,    15,    16],
        [17,    18,    19,    20],
        [21,    22,    23,    24],
    ]
]
, dtype="int32"
)
```

我们使用 tf.constant 创建了一个 2×3×4 的张量，数据类型为 int32，我们看到的数字是 1,2,3,…,24，我们将它分配给张量 tf_initial_tensor_constant。

现在打印出 tf_initial_tensor_constant 的 Python 变量来查看其所拥有的内容。

```
print(tf_initial_tensor_constant)
```

输出：

```
Tensor("Const:0", shape=(2, 3, 4), dtype=int32)
```

可以看到它是 TensorFlow 常量，形状是 2×3×4，数据类型是 int32。

因为我们还没有在 TensorFlow 会话中运行它，所以即使我们将它定义为常量，它似乎也没有值。这同样适用于我们即将创建的其他 reshape 张量。

第一个例子，我们将形状为 2×3×4 的张量更改为形状为 2×12 的张量。

```
tf_ex_one_reshaped_tensor_2_by_12 = tf.reshape(tf_initial_tensor_constant, [2, 12])
```

这里使用函数 tf.reshape()，传入 tf_initial_tensor_constant，还有想要的新形状的细节。新形状是 2、12，然后我们将整个事物分配给张量 tf_ex_one_reshaped_tensor_2_by_12。

请注意，元素的数量将保持不变，2×3×4 是 24，2×12 也是 24。

打印出张量 tf_ex_one_reshaped_tensor_2_by_12 来看看它有什么。

```
print(tf_ex_one_reshaped_tensor_2_by_12)
```

输出：

```
Tensor("Reshape:0", shape=(2, 12), dtype=int32)
```

我们看到它是 TensorFlow 张量，形状是 2×12，数据类型是 int32。

输出还没有显示任何值，因为我们仍在构建 TensorFlow 图，我们还没有在 TensorFlow 会话中运行它。

对于第二个例子，我们将形状为 2×3×4 的张量更改为形状为 2×3×2×2 的张量。

```
tf_ex_two_reshaped_tensor_2_by_3_by_2_by_2 = tf.reshape(tf_initial_tensor_constant, [2, 3, 2, 2])
```

这里使用函数 tf.reshape()，传入初始张量，指定目标形状。

所以传入 2、3、2、2 并将它分配给张量 tf_ex_two_reshaped_tensor_2_by_3_by_2_by_2。

请注意，元素的数量将保持不变，2×3×4 是 24，2×3×2×2 也是 24。

打印出张量 tf_ex_two_reshaped_tensor_2_by_3_by_2_by_2 来看看其有什么。

```
print(tf_ex_two_reshaped_tensor_2_by_3_by_2_by_2)
```

输出：

```
Tensor("Reshape_1:0", shape=(2, 3, 2, 2), dtype=int32)
```

我们看到它是 TensorFlow 张量，形状是 2×3×2×2，这是我们所期望的，数据类型是 int32。

对于第三个例子，我们将把形状为 2×3×4 的 TensorFlow 张量更改为 24 个元素的向量。

```
tf_ex_tre_reshaped_tensor_1_by_24 = tf.reshape(tf_initial_tensor_constant, [-1])
```

这里调用函数 tf.reshape()，传入初始张量，并使用参数[-1]。它的作用是将张量变平，所以得到一个包含 24 个元素的列表。我们将它分配给张量 tf_ex_tre_reshaped_tensor_1_by_24。

打印出张量 tf_ex_tre_reshaped_tensor_1_by_24，看看得到了什么。

```
print(tf_ex_tre_reshaped_tensor_1_by_24)
```

输出：

```
Tensor("Reshape_2:0", shape=(24,), dtype=int32)
```

我们看到它是 TensorFlow 张量，形状是 (24,)，这意味着它将是一个向量，数据类型是 int32。

现在我们已经创建了 TensorFlow 张量，是时候运行计算图了。首先，在会话中启动计算图。

```
sess = tf.Session()
```

然后初始化图中的所有全局变量。

```
sess.run(tf.global_variables_initializer())
```

在我们的例子中，它将是我们创建的所有张量。

接下来，我们将打印出 4 个张量，以了解 tf.reshape 的工作原理。

打印出我们的初始张量常数。

```
print(sess.run(tf_initial_tensor_constant))
```

输出：

```
[[[ 1   2   3   4]
  [ 5   6   7   8]
  [ 9  10  11  12]]

 [[13  14  15  16]
  [17  18  19  20]
```

```
    [21  22  23  24]]]
```

我们看到它是一个 2×3×4 张量，数字从 1 到 24，并且没有一个有小数点，所以我们知道它们是 int32 数字。

现在打印出我们的第一个重塑张量。

```
print(sess.run(tf_ex_one_reshaped_tensor_2_by_12))
```

输出：

```
[[ 1  2  3  4  5  6  7  8  9 10 11 12]
 [13 14 15 16 17 18 19 20 21 22 23 24]]
```

我们看到它是一个张量，里面有两个矩阵：第一个矩阵有 1 行和 12 列，第二个矩阵也有 1 行和 12 列，所有的元素 1～24 都在那里。

现在打印我们的第二个重塑张量 tf_ex_two_reshaped_tensor_2_by_3_by_2_by_2。

```
print(sess.run(tf_ex_two_reshaped_tensor_2_by_3_by_2_by_2))
```

输出：

```
[[[[ 1  2]
   [ 3  4]]

  [[ 5  6]
   [ 7  8]]

  [[ 9 10]
   [11 12]]]


 [[[13 14]
   [15 16]]

  [[17 18]
   [19 20]]

  [[21 22]
   [23 24]]]]
```

我们看到它是一个具有两个内部张量的张量，每个内部张量有 3 个 2×2 的矩阵。

所以是：2 行，2 列；2 行，2 列；2 行，2 列。然后 2 行，2 列；2 行，2 列；2 行，2 列。

总的来说，我们可以看到形状是 2×3×2×2，所有的数字都在那里。

最后，打印我们的第三个重塑的 TensorFlow 例子。

```
print(sess.run(tf_ex_tre_reshaped_tensor_1_by_24))
```

Python 变量 tf_ex_tre_reshaped_tensor_1_by_24 的值如下：

```
[ 1  2  3  4  5  6  7  8  9 10 11 12 13 14 15 16 17 18 19 20 21 22 23 24]
```

我们看到它是一个 24 个元素长的向量。所以我们看到数字从 1 一直到 24。只要元素的数量保持不变，我们就可以使用函数 tf.reshape() 来改变 TensorFlow 张量的形状。

现在，我们拥有了所有必需的材料，可以开始构建带有一个隐藏层和 200 个隐藏单元（神经元）的前馈神经网络。TensorFlow 中的计算图如图 3-8 所示。

在现实世界的问题中，我们有成千上万的输入，这使得梯度下降的计算成本很高。这就是为什么我们将输入集分成几个尺寸为 B（称为小批量尺寸）的较短片段（称为小批量），并逐个输入它们。这称为"随机梯度下降"(Stochastic Gradient Descent)。将尺寸大小为 B 的每个小批量馈送到网络，反向传播错误以及更新参数（权重和偏差）的过程称为迭代。

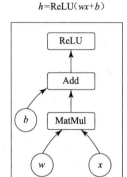

图 3-8　h 的计算图

我们通常使用占位符来输入，以便可以在上下文中没有任何实际值的情况下构建图。唯一的一点是需要为输入选择适当的尺寸。在这里，我们有一个前馈神经网络，我们假设尺寸为 784 的输入（类似于 MNIST 数据的 28×28 图像）。输入占位符可以写为：

```
#创建输入占位符
x = tf.placeholder(tf.float32, shape=[None, 784], name="x")
```

你可能想知道为什么是 shape=[None, 784]？因为我们需要在每次训练迭代中将 B 个尺寸为 784 的图像作为一个批次提供给网络。所以占位符需要是 shape = [B, 784]。将占位符的形状定义为 [None, 784] 意味着我们可以提供任何数量的大小为 784 的图像（不一定是 B 个图像）。这在评估时特别有用，我们需要将所有验证或测试图像提供给网络并计算所有验证或测试图像的性能。

接下来，我们来看看网络参数 w 和 b。正如上面的变量部分所解释的那样，它们必须被定义为变量。由于在 TensorFlow 中，默认情况下，渐变更新将应用于图形变量。如前所述，变量需要初始化。

一般来说，权重 (w) 是随机初始化的，它是正态分布中最简单的形式，比如零均值和标准差为 0.01 的正态分布。偏差 (b) 可以初始化为小的常数值，例如 0。

由于输入维数为 784，并且我们有 200 个隐藏单位，所以权重矩阵的大小为[784,200]。我们还需要 200 个偏差，每个隐藏单位一个。代码如下所示：

```
#创建从 N(0, 0.01)随机初始化的权重矩阵
weight_initer = tf.truncated_normal_initializer(mean=0.0, stddev=0.01)
```

```
W = tf.get_variable(name="Weight", dtype=tf.float32, \
   shape=[784, 200], initializer=weight_initer)

#创建尺寸为 200 的偏置矢量,全部初始化为零
bias_initer =tf.constant(0., shape=[200], dtype=tf.float32)
b = tf.get_variable(name="Bias", dtype=tf.float32, initializer=bias_initer)
```

现在来看看矩阵运算。我们必须乘以输入 $X_{[None,784]}$ 和权重矩阵 $W_{[784,200]}$,这个运算给出尺寸为 [None,200] 的张量,然后加上偏差向量 $b_{[200]}$,并最后从一个 ReLU 非线性产生最终张量。

```
#创建 MatMul 节点
x_w = tf.matmul(X, W, name="MatMul")
#创建 Add 节点
x_w_b = tf.add(x_w, b, name="Add")
#创建 ReLU 节点
h = tf.nn.relu(x_w_b, name="ReLU")
```

在关闭它之前,让我们在该图上运行会话(使用由随机像素值生成的 100 张图像)并获取隐藏单元的输出(h)。以下是完整的代码:

```
import tensorflow as tf
import numpy as np

#创建输入占位符
X = tf.placeholder(tf.float32, shape=[None, 784], name="X")
weight_initer = tf.truncated_normal_initializer(mean=0.0, stddev=0.01)

#创建网络参数
W = tf.get_variable(name="Weight", dtype=tf.float32, \
   shape=[784, 200], initializer=weight_initer)
bias_initer =tf.constant(0., shape=[200], dtype=tf.float32)
b = tf.get_variable(name="Bias", dtype=tf.float32, initializer=bias_initer)

#创建 MatMul 节点
x_w = tf.matmul(X, W, name="MatMul")
#创建 Add 节点
x_w_b = tf.add(x_w, b, name="Add")
#创建 ReLU 节点
h = tf.nn.relu(x_w_b, name="ReLU")

#添加一个运算来初始化全局变量
init_op = tf.global_variables_initializer()

#在会话中运行计算图
with tf.Session() as sess:
    #初始化变量
```

```
sess.run(init_op)
#创建词典
d = {X: np.random.rand(100, 784)}
#通过词典将值提供给占位符
print(sess.run(h, feed_dict=d))
```

TensorFlow 计算图虽然功能强大，但可能变得非常复杂。使用 TensorBoard 可视化图形可以帮助我们理解和调试计算图。

为了让我们的 TensorFlow 程序激活 TensorBoard，需要添加一些代码行。这会将 TensorFlow 运算导出到称为事件文件（或事件日志文件）的文件中。TensorBoard 能够读取此文件并提供模型图形及其性能的一些可视化显示。

现在让我们编写一个简单的 TensorFlow 程序，并用 TensorBoard 可视化其计算图。

创建两个常量并将它们加在一起。常量张量可以简单地由它们的值来定义：

```
import tensorflow as tf

#创建计算图
a = tf.constant(2)
b = tf.constant(3)
c = tf.add(a, b)
#在会话中运行计算图
with tf.Session() as sess:
    print(sess.run(c))
```

为了用 TensorBoard 把程序可视化，需要编写程序的日志文件。要编写事件日志文件，首先需要使用以下代码为这些日志创建一个写入器：

```
writer = tf.summary.FileWriter([logdir], [graph])
```

其中[logdir]是想要存储这些日志文件的文件夹。也可以选择[logdir]为 "./graphs" 等有意义的名字。第二个参数[graph]是正在开发的程序的图形。有两种方法可以获得图形：

（1）使用 tf.get_default_graph()调用图形，该函数返回程序的默认图形。

（2）将其设置为返回会话的图形的 sess.graph（请注意，这需要创建一个会话）。

下面的例子让我们两种方式都看到。然而，第二种方式更常见。无论哪种方式，确保在定义图形后才能创建一个写入器；否则，在 TensorBoard 上可视化的图形将不完整。

将写入器添加到第一个示例中，并将图形可视化。

```
import tensorflow as tf
tf.reset_default_graph()    #清除先前单元格的已定义变量和运算

#创建图形
a = tf.constant(2)
```

```
b = tf.constant(3)
c = tf.add(a, b)

#在会话外创建写入器
#writer = tf.summary.FileWriter('./graphs', tf.get_default_graph())

#在会话中启动图形
with tf.Session() as sess:
    #或在会话中创建写入器
    writer = tf.summary.FileWriter('./graphs', sess.graph)
    print(sess.run(c))
```

现在，如果执行此代码，TensorFlow 将在当前目录中创建一个包含事件文件的目录。

接下来，在运行 Python 代码的目录下启动 TensorBoard 服务：

```
$ tensorboard --logdir="./graphs" --port 6006
```

然后在浏览器中使用 http://<IP_Address>:6006/访问。

该链接将引导我们进入 TensorBoard 页面，它应该类似于图 3-9。

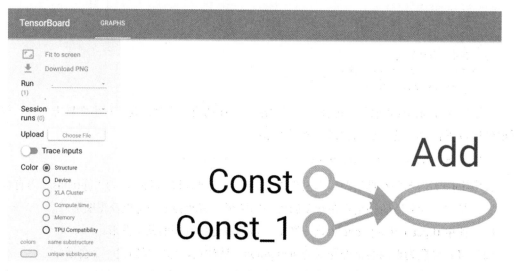

图 3-9　TensorBoard 页面可视化示例代码中生成的图形

图 3-9 显示了模型的各个部分。图 3-9 中的 Const 和 Const_1 对应于 a 和 b，节点 "Add" 对应于 c。代码中给出的名称（a、b 和 c）只是它们在 Python 中的名称，它们只在编写代码时帮助我们访问。名称对 TensorFlow 和 TensorBoard 没有任何意义。为了使 TensorBoard 了解运算的名称，必须明确地命名它们。

再次修改代码来添加名称：

```python
import tensorflow as tf
tf.reset_default_graph()    #清除先前单元格的已定义变量和运算

#创建图形
a = tf.constant(2, name="a")
b = tf.constant(3, name="b")
c = tf.add(a, b, name="addition")

#在会话外创建写入器
#writer = tf.summary.FileWriter('./graphs', tf.get_default_graph())

#在会话中启动图形
with tf.Session() as sess:
    #或在会话中创建写入器
    writer = tf.summary.FileWriter('./graphs', sess.graph)
    print(sess.run(c))
```

产生的页面如图 3-10 所示。

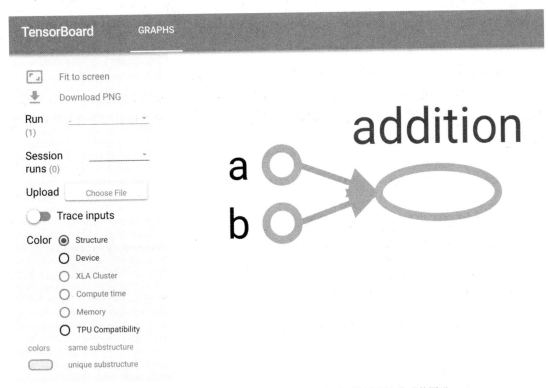

图 3-10　TensorBoard 页面使用修改后的名称可视化示例代码中生成的图形

到目前为止，我们只关注如何在 TensorBoard 中可视化图形。TensorBoard 还提供其他类型的可视化（标量、图像和直方图）。在这一部分，我们将使用一种称为摘要的特殊操作来将模型参数（如神经网络的权重和偏差）、度量（如损失或准确度值）和图像（如到网络的输入图像）可视化。

摘要（Summary）是一个特殊的运算，它接收常规张量并将汇总数据输出到磁盘（即事件文件中）。

基本上，有以下3种主要类型的摘要。

（1）tf.summary.scalar：用于写入单个标量值张量（如分类损失或准确度值）。

（2）tf.summary.histogram：用于绘制非标量张量的所有值的直方图（可用于可视化神经网络的权重或偏差矩阵）。

（3）tf.summary.image：用于绘制图像（如网络的输入图像或自动编码器或 GAN 生成的输出图像）。

接下来，将更详细地介绍上述所有摘要类型。

tf.summary.scalar 用于编写随时间或迭代而变化的标量张量的值。在神经网络（例如分类任务的简单网络）中，通常用于监视损失函数或分类精度的变化。

让我们举一个简单的例子来理解这一点。

从标准正态分布 $N(0,1)$ 中随机挑选 100 个值，并依次绘制它们。

一种方法是简单地创建一个变量，并从正态分布（平均值为 0 和标准差为 1）初始化变量，然后在会话中运行 for 循环并初始化它 100 次。代码如下所示，编写摘要所需的步骤在代码中进行了说明。

```
import tensorflow as tf
tf.reset_default_graph()    #清除先前单元格的已定义变量和运算

#创建标量变量
x_scalar = tf.get_variable('x_scalar',  \
    shape=[], initializer=tf.truncated_normal_initializer(mean=0, stddev=1))

#_____步骤1:_____创建标量摘要
first_summary = tf.summary.scalar(name='My_first_scalar_summary', tensor=x_scalar)

init = tf.global_variables_initializer()

#在会话中启动图形
with tf.Session() as sess:
    #_____步骤2:_____在会话中创建写入器
    writer = tf.summary.FileWriter('./graphs', sess.graph)
    for step in range(100):
        #循环变量的初始化
```

```
        sess.run(init)
        #____步骤 3：____评估标量摘要
        summary = sess.run(first_summary)
        #____步骤 4：____将摘要添加到写入器（即事件文件）以写入硬盘
        writer.add_summary(summary, step)
    print('Done with writing the scalar summary')
```

如果希望观察值随时间或迭代的变化，则直方图会派上用场。它用于绘制非标量张量值的直方图。这为我们提供了张量值的直方图（和分布）如何随着时间或迭代而变化的视图。

在神经网络的情况下，它通常用于监测权重和偏差分布的变化。它对于检测网络参数的不规则行为非常有用（例如，当权重发生爆炸或异常收缩时）。

现在让我们回到前面的例子，并将直方图摘要添加到那个例子。

继续前面的示例，添加一个大小为 30×40 的矩阵，其条目来自标准正态分布。 初始化该矩阵 100 次，并绘制其输入项随时间的分布。

```
import tensorflow as tf
tf.reset_default_graph()    #清除先前单元格的已定义变量和运算

#创建变量
x_scalar = tf.get_variable('x_scalar', \
 shape=[], initializer=tf.truncated_normal_initializer(mean=0, stddev=1))
x_matrix = tf.get_variable('x_matrix', \
   shape=[30, 40], initializer=tf.truncated_normal_initializer(mean=0, stddev=1))

#____步骤 1：____创建摘要
#标量张量的标量摘要
scalar_summary = tf.summary.scalar('My_scalar_summary', x_scalar)
#非标量（即 2D 或矩阵）张量的直方图摘要
histogram_summary = tf.summary.histogram('My_histogram_summary', x_matrix)

init = tf.global_variables_initializer()

#在会话中启动图形
with tf.Session() as sess:
    #____步骤 2：____在会话中创建写入器
    writer = tf.summary.FileWriter('./graphs', sess.graph)
    for step in range(100):
        #循环变量的几个初始化运算
        sess.run(init)
        #____步骤 3：____评估合并的摘要
        summary1, summary2 = sess.run([scalar_summary, histogram_summary])
        #____步骤 4：____将摘要添加到写入器（即事件文件）以写入硬盘
        writer.add_summary(summary1, step)
```

```
#重复直方图摘要的步骤 4
    writer.add_summary(summary2, step)
print('Done writing the summaries')
```

在 TensorBoard 中,顶部菜单中添加了两个新选项卡:"分布"和"直方图",结果如图 3-11 所示。

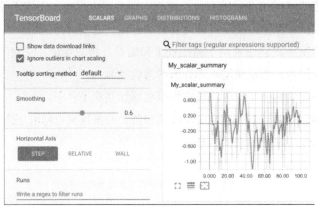

(a)标量摘要

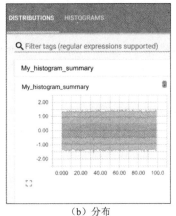

(b)分布

(c)直方图

图 3-11 "分布"和"直方图"选项卡及二维张量值超过 100 步的直方图

在该图中,"分布"(Distributions)选项卡包含一个图表,显示通过步骤(x 轴)与张量值(y 轴)的分布。

我们需要运行每个摘要(例如 sess.run([scalar_summary, histogram_summary])),然后使用写入器将它们中的每一个写入磁盘。实际上,可以使用任意数量的摘要来跟踪模型中的不同参数。这使得运行和写入摘要极其低效。解决方法是合并图表中的所有摘要,并在会话中立即运行它们。这可以通过 tf.summary.merge_all()方法来完成。让我们将其添加到示例 3,代码更改

如下:

```python
import tensorflow as tf
tf.reset_default_graph()   #清除先前单元格的已定义变量和运算

#创建变量
x_scalar = tf.get_variable('x_scalar', \
    shape=[], initializer=tf.truncated_normal_initializer(mean=0, stddev=1))
x_matrix = tf.get_variable('x_matrix', \
    shape=[30, 40], initializer=tf.truncated_normal_initializer(mean=0, stddev=1))

#____步骤1:____创建摘要
#标量张量的标量摘要
scalar_summary = tf.summary.scalar('My_scalar_summary', x_scalar)
#非标量(即2D或矩阵)张量的直方图摘要
histogram_summary = tf.summary.histogram('My_histogram_summary', x_matrix)

#____步骤2:____合并所有摘要
merged = tf.summary.merge_all()

init = tf.global_variables_initializer()

#在会话中启动图形
with tf.Session() as sess:
    #____步骤3:____在会话中创建写入器
    writer = tf.summary.FileWriter('./graphs', sess.graph)
    for step in range(100):
        #循环变量的几个初始化运算
        sess.run(init)
        #____步骤4:____评估合并的摘要
        summary = sess.run(merged)
        #____步骤5:____将摘要添加到写入器(即事件文件)以写入硬盘
        writer.add_summary(summary, step)
    print('Done writing the summaries')
```

tf.summary.image 用于写出和将可视化张量作为图像。在神经网络的情况下,这通常用于追踪馈送到网络(例如在每批中)或者在输出中生成的图像(例如在自动编码器中重建的图像)。但是,一般来说,这可以用于绘制任何张量。例如,可以将尺寸为 30×40 的权重矩阵可视化为 30×40 像素的图像。

图像摘要可以使用以下代码创建:

```
tf.summary.image(name, tensor, max_outputs=3)
```

其中 name 是生成节点的名称(即运算),tensor 是要写成图像摘要的期望张量,max_outputs

是张量中生成图像的最大元素数目。但是，这是什么意思？答案在于张量的形状。

我们提供给 tf.summary.image 的张量必须是形状为四维的张量[batch_size, height, width, channels]，其中 batch_size 是批次中图像的数量，height 和 width 决定图像的大小。最后，channels 的值包括 1、3、4。其中，1 用于灰度图像；3 用于 RGB（即彩色）图像；4 用于 RGBA 图像（其中 A 代表 alpha 值）。

下面来看一个非常简单的例子。

```
import tensorflow as tf
tf.reset_default_graph()   #清除先前单元格的已定义变量和运算

#创建变量
w_gs = tf.get_variable('W_Grayscale', \
    shape=[30, 10], initializer=tf.truncated_normal_initializer(mean=0, stddev=1))
w_c = tf.get_variable('W_Color', \
    shape=[50, 30], initializer=tf.truncated_normal_initializer(mean=0, stddev=1))

#_____步骤 0:_____将变量重新塑造成四维张量
w_gs_reshaped = tf.reshape(w_gs, (3, 10, 10, 1))
w_c_reshaped = tf.reshape(w_c, (5, 10, 10, 3))

#_____步骤 1:_____创建摘要
gs_summary = tf.summary.image('Grayscale', w_gs_reshaped)
c_summary = tf.summary.image('Color', w_c_reshaped, max_outputs=5)

#_____步骤 2:_____合并所有摘要
merged = tf.summary.merge_all()

#创建用于初始化所有变量的运算
init = tf.global_variables_initializer()

#在会话中启动图形
with tf.Session() as sess:
    #_____步骤 3:_____在会话中创建写入器
    writer = tf.summary.FileWriter('./graphs', sess.graph)
    #初始化所有变量
    sess.run(init)
    #_____步骤 4:_____评估合并的运算以获得摘要
    summary = sess.run(merged)
    #_____步骤 5:_____将摘要添加到写入器（即事件文件）以写入硬盘
    writer.add_summary(summary)
    print('Done writing the summaries')
```

现在像以前一样打开 TensorBoard 并切换到 IMAGES 选项卡。图像应该类似于图 3-12。

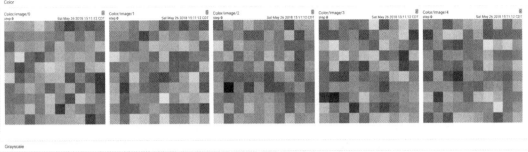

图 3-12 在 TensorBoard 中生成的图像

可以类似地将任何尺寸的其他图像添加到摘要中,并将它们绘制在 TensorBoard 中。

下面讨论如何将参数保存到磁盘并从磁盘恢复已保存的参数。网络的可保存/可恢复的参数是变量(即权重和偏差)。

为了保存和恢复变量,所需要做的就是在图结尾调用 tf.train.Saver()。

```
#创建图
X = tf.placeholder(..)
Y = tf.placeholder(..)
w = tf.get_variable(..)
b = tf.get_variable(..)
...
loss = tf.losses.mean_squared_error(..)
optimizer = tf.train.AdamOptimizer(..).minimize(loss)
...

saver = tf.train.Saver()
```

训练模式下,在会话中我们将初始化变量并运行网络。在训练结束时,将使用 saver.save() 保存变量:

```
#训练
with tf.Session() as sess:
    sess.run(tf.global_variables_initializer())
    #训练模型
    for step in range(steps):
```

```
        sess.run(optimizer)
        ...
    saved_path = saver.save(sess, './my-model', global_step=step)
```

这将创建 3 个文件（.data、.index、.meta），并带有所保存模型的步骤的后缀。

测试模式下，在会话中我们将使用 saver.restore()恢复变量并验证或测试模型。

```
#测试
with tf.Session() as sess:
    saver.restore(sess, './my-model')
    ...
```

实现在 TensorFlow 中保存和恢复两个变量。我们将创建一个包含两个变量的图。我们来创建两个变量 a = [3 3]和 b = [5 5 5]：

```
import tensorflow as tf
#创建变量a和b
a = tf.get_variable("A", initializer=tf.constant(3, shape=[2]))
b = tf.get_variable("B", initializer=tf.constant(5, shape=[3]))
```

请注意小写字母为变量的 Python 名称，大写字母为张量的 TensorFlow 名称。当我们想要导入图来恢复数据时，这一点很重要。

变量在使用前需要初始化。为此，必须调用变量初始值设定运算并在会话中运行该运算。这是一次初始化所有变量的最简单方法。

```
#初始化所有变量
init_op = tf.global_variables_initializer()
```

现在，在会话中，我们可以初始化变量并运行以查看值：

```
#运行会话
with tf.Session() as sess:
    #初始化会话中的所有变量
    sess.run(init_op)
    #运行会话以获取变量的值
    a_out, b_out = sess.run([a, b])
    print('a = ', a_out)
    print('b = ', b_out)
```

输出：

```
a = [3 3]
b = [5 5 5]
```

所有变量都存在于会话范围内。所以，在会话结束后，我们将放弃这些变量。

为了保存变量，我们将在图中使用 tf.train.Saver()调用保存函数。该函数将查找图中的所有变量。我们可以在_var_list 中看到所有变量的列表。下面创建一个 saver 对象并查看对象中的

_var_list：

```
#创建 saver 对象
saver = tf.train.Saver()
for i, var in enumerate(saver._var_list):
    print('Var {}: {}'.format(i, var))
```

运行后输出：

```
Var 0: <tf.Variable 'A:0' shape=(2,) dtype=int32_ref>
Var 1: <tf.Variable 'B:0' shape=(3,) dtype=int32_ref>
```

所以，我们的图由上面列出的两个变量组成。请注意：变量名称末尾有 0。

既然保存对象是在图形中创建的，那么在会话中，就可以调用 saver.save()函数将变量保存在磁盘中。我们必须将创建的会话（sess）和想要保存变量的文件路径传递给函数 save()：

```
#运行会话
with tf.Session() as sess:
    #初始化会话中的所有变量
    sess.run(init_op)

    #将变量保存在磁盘中
    saved_path = saver.save(sess, './saved_variable')
    print('model saved in {}'.format(saved_path))
```

模型保存在./saved_variable 中。

如果检查自己的工作目录，会注意到已经创建了 3 个名为 saved_variable 的新文件。

```
import os
for file in os.listdir('.'):
    if 'saved_variable' in file:
        print(file)
```

输出：

```
saved_variable.data-00000-of-00001
saved_variable.meta
saved_variable.index
```

这里的.data 文件包含变量值；.meta 文件包含图形结构；.index 文件标识检查点。

现在，所需要的所有内容都已保存在磁盘中，可以使用 saver.restore()在会话中加载已保存的变量：

```
#运行会话
with tf.Session() as sess:
    #恢复保存的变量
    saver.restore(sess, './saved_variable')
```

```
#打印加载的变量
a_out, b_out = sess.run([a, b])
print('a = ', a_out)
print('b = ', b_out)
```

运行后输出结果：

```
INFO:tensorflow:Restoring parameters from ./saved_variable
a = [3 3]
b = [5 5 5]
```

请注意，这次我们没有在会话中初始化变量；相反，是从磁盘中恢复它们。

重要提示：为了恢复参数，应该定义图形。由于我们在顶部定义了图形，因此在恢复参数时没有问题。

除了在代码中定义图形，还可以从 .meta 文件恢复图形。当保存这些变量时，它会创建一个 .meta 文件。该文件包含图形结构。因此，可以使用 tf.train.import_meta_graph() 导入元图并恢复图的值。导入图形并查看图中的所有张量：

```
#删除当前图形
tf.reset_default_graph()

#从文件中导入图形
imported_graph = tf.train.import_meta_graph('saved_variable.meta')

#列出图中的所有张量
for tensor in tf.get_default_graph().get_operations():
    print (tensor.name)
```

现在我们有导入的图形，并且我们对张量 A 和 B 感兴趣，因此我们可以恢复参数：

```
#运行会话
with tf.Session() as sess:
    #恢复保存的张量
    imported_graph.restore(sess, './saved_variable')
    #打印加载的张量
    a_out, b_out = sess.run(['A:0','B:0'])
    print('a = ', a_out)
    print('b = ', b_out)
```

运行输出结果如下：

```
INFO:tensorflow:Restoring parameters from ./saved_variable
a = [3 3]
b = [5 5 5]
```

请注意，在 sess.run() 中，我们提供张量的 TensorFlow 名称是 'A:0' 和 'B:0'，而不是 a 和 b。

Saver 主要用于生成变量的检查点。SavedModel 将取代现有的 TensorFlow 推理模型格式，作为导出 TensorFlow 图形进行服务的标准方式。

TensorFlow 的 SavedModel 格式包括有关模型（图形、检查点状态和其他元数据）的所有信息。所以，如果想在 Java 中使用，可以使用 SavedModelBundle.load 创建使用训练变量值初始化的会话。

Python 中的保存模型代码如下：

```python
import tensorflow as tf
import os

x = tf.placeholder(tf.float32, name='x')                    #模型输入
w = tf.get_variable('w', shape=[1, 1], initializer=tf.random_normal_initializer())
b = tf.get_variable('b', shape=[1, 1], initializer=tf.zeros_initializer())
y = tf.add(b, tf.matmul(x, w), name='y')                    #模型输出

training_steps = 51

with tf.Session() as sess:
    sess.run(tf.global_variables_initializer())

    for train_step in range(1, training_steps):
        #...
        #运行训练运算
        #...
        if train_step % 50 == 0:                            #每隔50步保存一次
            #将 train_step 添加到 save_dir
            export_dir = os.getcwd() + '/savedModel-' + str(train_step)
            inputs_dict = {'x_in': x}                       #字符串名称不等于张量名称
            outputs_dict = {'y': y}                         #字符串名称等于张量名称
            tf.saved_model.simple_save(sess, export_dir, inputs=inputs_dict, outputs=outputs_dict)

            print('y: %.4f' % (sess.run(y, feed_dict={ x: [[1]] })))


tf.reset_default_graph()         #清除图结构

with tf.Session() as sess:

    #加载图形和变量
    export_dir = os.getcwd() + '/savedModel-50'
    meta_graph_def = tf.saved_model.loader.load(sess, \
        [tf.saved_model.tag_constants.SERVING], export_dir)
```

```
    #获取将输入/输出字符串名称映射到张量的签名定义
sig_def = \
 meta_graph_def.signature_def[ \
tf.saved_model.signature_constants.DEFAULT_SERVING_SIGNATURE_DEF_KEY]

    x_name = sig_def.inputs['x_in'].name          #获取输入张量名称
    x = tf.get_default_graph().get_tensor_by_name(x_name)

    y_name = sig_def.outputs['y'].name            #获取输出张量名称
    y = tf.get_default_graph().get_tensor_by_name(y_name)

    print('y: %.4f' % (sess.run(y, feed_dict={ x: [[1]] })))

#Output:
#x: -0.2909
#x: -0.2909
```

训练时执行的计算过程和推理时执行的计算过程不一样。

神经网络是深度学习的第一步。深度学习的名字来源于计算机科学家希望用神经元的相同功能来模拟大脑结构的概念。深度学习的关键特点是它可以分离不能线性分离的数据。

要构建任何分类器，您的代码需要如下部分：

（1）为网络准备所需的库，输入数据和超参数。

（2）建立网络图。

（3）训练网络。

（4）测试网络。

为了能够使用 matplotlib，要首先安装 matplotlib 依赖的 tkinter：

```
$ sudo apt-get install python3-tk
```

从导入所需的库开始。

```
#imports
import tensorflow as tf
import numpy as np
import matplotlib.pyplot as plt
```

在这里，我们使用 MNIST 数据集。MNIST 是手写数字的数据集。MNIST 是深度学习数据集的基准。我们使用 MNIST 的另一个原因是通过 Tensorflow 很容易访问。

该数据集包含 55000 个训练示例，其中 5000 个是用于验证的示例，10000 个是用于测试的示例。这些数字图像已经进行了尺寸标准化并以固定尺寸图像（28×28 像素）为中心。图像中的像素点用值为 0 或 1 的数表示。为了简单起见，每幅图像都被展平并转换为具有 784 个特征的一维 numpy 阵列（28×28）。

我们可以轻松导入数据集并查看训练、测试和验证集的大小：

```
#导入 MNIST 数据
from tensorflow.examples.tutorials.mnist import input_data
mnist = input_data.read_data_sets("MNIST_data/", one_hot=True)

print("Size of:")
print("- Training-set:\t\t{}".format(len(mnist.train.labels)))
print("- Test-set:\t\t{}".format(len(mnist.test.labels)))
print("- Validation-set:\t{}".format(len(mnist.validation.labels)))
```

运行结果：

```
Size of:
- Training-set:         55000
- Test-set:             10000
- Validation-set:       5000
```

超参数是网络无法学习的重要参数，所以必须在外部指定超参数。

```
#超参数
learning_rate = 0.001    #优化学习率
epochs = 10              #训练回合数
batch_size = 100         #训练批次大小
display_freq = 100       #显示训练结果的频率

#网络参数
#MNIST 图像在每个维度上是 28 像素
img_h = img_w = 28

#图像存储在这个长度的一维数组中
img_size_flat = img_h * img_w

#类的数量，10 个数字每个数字一个类
n_classes = 10

#第一个隐藏层中的单元数量
h1 = 200
```

在我们开始构建图之前，需要快速地完成一些功能。因此，我们不会多次调用它们，而是定义一些有用的函数，并会在图中调用这些函数。最重要的是用于创建权重和偏差变量的函数。由于我们正在创建一个神经网络，因此需要一个完全连接的层来将上一层的所有节点连接到我们的层。

```
#权重和偏差包装器，可以用于设置卷积核的权重
def weight_variable(name, shape):
    """
```

```python
    用适当的初始化创建一个权重变量
    name：权重变量名称
    shape：权重变量形状
    return：初始化的权重变量
    """
    initer = tf.truncated_normal_initializer(stddev=0.01)
    return tf.get_variable('W_' + name,
                           dtype=tf.float32,
                           shape=shape,
                           initializer=initer)


def bias_variable(name, shape):
    """
    用适当的初始化创建一个偏差变量
    name：偏差变量名称
    shape：偏差变量形状
    return：初始化的偏差变量
    """
    initial = tf.constant(0., shape=shape, dtype=tf.float32)
    return tf.get_variable('b_' + name,
                           dtype=tf.float32,
                           initializer=initial)


def fc_layer(x, num_nodes, name, use_relu=True):
    """
    创建一个完全连接的图层
    :param x：来自上一层的输入
    :param num_nodes：完全连接层中隐藏单元的数量
    :param name：图层名称
    :param use_relu：用来决定是否添加 ReLU 非线性到图层的布尔值

    :return：输出数组
    """
    in_dim = x.get_shape()[1]
    W = weight_variable(name, shape=[in_dim, num_nodes])
    b = bias_variable(name, [num_nodes])
    layer = tf.matmul(x, W)
    layer += b
    if use_relu:
        layer = tf.nn.relu(layer)

    return layer
```

既然有了帮助函数，即可以创建我们的图结构：

```
#创建图结构
#inputs (x)和outputs(y)的占位符
x = tf.placeholder(tf.float32, shape=[None, img_size_flat], name='X')
y = tf.placeholder(tf.float32, shape=[None, n_classes], name='Y')
fc1 = fc_layer(x, h1, 'FC1', use_relu=True)
output_logits = fc_layer(fc1, n_classes, 'OUT', use_relu=False)

#定义损失函数，优化器和准确度
loss = \
    tf.reduce_mean(tf.nn.softmax_cross_entropy_with_logits(labels=y, \
            logits=output_logits), \
            name='loss')
optimizer = \
 tf.train.AdamOptimizer(learning_rate=learning_rate, \
    name='Adam-op').minimize(loss)
correct_prediction = \
 tf.equal(tf.argmax(output_logits, 1), tf.argmax(y, 1), name='correct_pred')
accuracy = tf.reduce_mean(tf.cast(correct_prediction, tf.float32), name='accuracy')

#网络预测
cls_prediction = tf.argmax(output_logits, axis=1, name='predictions')

#初始化变量
init = tf.global_variables_initializer()
```

一旦创建出图结构以后，就可以在一个会话上运行它。运行时可以使用tf.Session()。但是，一旦单元运行，会话就会结束，我们将丢失所有信息。所以，我们将定义一个InteractiveSession来保存参数用于测试。

```
#在会话中启动图形
sess = tf.InteractiveSession() #使用InteractiveSession而不是Session来测试单独单元中
                               #的网络
sess.run(init)
#每个回合的训练迭代次数
num_tr_iter = int(mnist.train.num_examples / batch_size)
for epoch in range(epochs):
    print('Training epoch: {}'.format(epoch+1))
    for iteration in range(num_tr_iter):
        batch_x, batch_y = mnist.train.next_batch(batch_size)

        #运行优化运算（反向传播）
        feed_dict_batch = {x: batch_x, y: batch_y}
        sess.run(optimizer, feed_dict=feed_dict_batch)
```

```python
        if iteration % display_freq == 0:
            #计算并显示批损失和准确度
            loss_batch, acc_batch = sess.run([loss, accuracy],
                                    feed_dict=feed_dict_batch)
            print("iter {0:3d}:\t Loss={1:.2f},\tTraining Accuracy={2:.01%}".
                format(iteration, loss_batch, acc_batch))

    #在每个回合后运行验证
    feed_dict_valid = {x: mnist.validation.images, y: mnist.validation.labels}
    loss_valid, acc_valid = sess.run([loss, accuracy], feed_dict=feed_dict_valid)
    print('---------------------------------------------------------')
    print("Epoch: {0}, validation loss: {1:.2f}, validation accuracy: {2:.01%}".
        format(epoch + 1, loss_valid, acc_valid))
    print('---------------------------------------------------------')
```

使用上述代码训练好模型后,现在是测试我们模型的时候了。我们将定义一些辅助函数来绘制一些图像及其相应的预测和真实类。我们还将可视化一些错误分类的样本,以了解为什么神经网络未能正确分类它们。

```python
def plot_images(images, cls_true, cls_pred=None, title=None):
    """
    用 3x3 子图创建图
    :param images: 要绘制的图像数组 (9, img_h*img_w)
    :param cls_true: 相应的真值标签(9,)
    :param cls_pred: 相应的真值标签(9,)
    """
    fig, axes = plt.subplots(3, 3, figsize=(9, 9))
    fig.subplots_adjust(hspace=0.3, wspace=0.3)
    img_h = img_w = np.sqrt(images.shape[-1]).astype(int)
    for i, ax in enumerate(axes.flat):
        #绘制图像
        ax.imshow(images[i].reshape((img_h, img_w)), cmap='binary')

        #显示实际和预测的类
        if cls_pred is None:
            ax_title = "True: {0}".format(cls_true[i])
        else:
            ax_title = "True: {0}, Pred: {1}".format(cls_true[i], cls_pred[i])

        ax.set_title(ax_title)

        #从图中删除刻度线
        ax.set_xticks([])
        ax.set_yticks([])
```

```python
    if title:
        plt.suptitle(title, size=20)
    plt.show()


def plot_example_errors(images, cls_true, cls_pred, title=None):
    """
    绘制错误分类图像示例的函数
    :param images: 所有图像的数组, (#imgs, img_h*img_w)
    :param cls_true: 相应的真实值标签, (#imgs,)
    :param cls_pred: 相应的预测标签, (#imgs,)
    """
    #否定布尔数组
    incorrect = np.logical_not(np.equal(cls_pred, cls_true))

    #从测试集中获取错误分类的图像
    incorrect_images = images[incorrect]

    #获取这些图像的真实和预测的类
    cls_pred = cls_pred[incorrect]
    cls_true = cls_true[incorrect]

    #绘制前9张图像
    plot_images(images=incorrect_images[0:9],
                cls_true=cls_true[0:9],
                cls_pred=cls_pred[0:9],
                title=title)
```

服务器版本的Ubuntu需要安装图形界面：

```
$ sudo apt install lubuntu-desktop
```

然后重启：

```
$ reboot
```

运行模型：

```python
#训练后测试网络
#测试模型的准确度
feed_dict_test = {x: mnist.test.images, y: mnist.test.labels}
loss_test, acc_test = sess.run([loss, accuracy], feed_dict=feed_dict_test)
print('---------------------------------------------------------')
print("Test loss: {0:.2f}, test accuracy: {1:.01%}".format(loss_test, acc_test))
print('---------------------------------------------------------')

#绘制正确和错误的分类例子
```

```
cls_pred = sess.run(cls_prediction, feed_dict=feed_dict_test)
cls_true = np.argmax(mnist.test.labels, axis=1)
plot_images(mnist.test.images, cls_true, cls_pred, title='Correct Examples')
plot_example_errors(mnist.test.images, cls_true, cls_pred, title='Misclassified Examples')
```

使用卷积神经网络可以减少分类错误。

接下来，我们在 TensorFlow 中实现一个简单的卷积神经网络。用两个卷积层，后接两个全连接层，如图 3-13 所示。

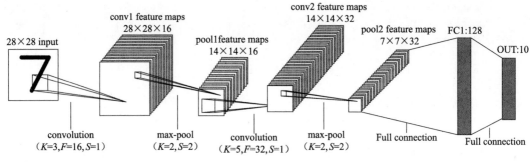

图 3-13　用于数字识别的 CNN 结构

加载 MNIST 数据的辅助函数：

```
def load_data(mode='train'):
    """
    下载和加载 MNIST 数据的函数
    :param mode: train 或 test
    :return: 图像和相应的标签
    """
    mnist = input_data.read_data_sets("MNIST_data/", one_hot=True)
    if mode == 'train':
        x_train, y_train, x_valid, y_valid = mnist.train.images, mnist.train.labels, \
                                             mnist.validation.images, mnist.validation.labels
        x_train, _ = reformat(x_train, y_train)
        x_valid, _ = reformat(x_valid, y_valid)
        return x_train, y_train, x_valid, y_valid
    elif mode == 'test':
        x_test, y_test = mnist.test.images, mnist.test.labels
        x_test, _ = reformat(x_test, y_test)
    return x_test, y_test


def reformat(x, y):
```

```python
"""
将数据重新格式化为卷积层可接受的格式
:param x: 输入数组
:param y: 相应的标签
:return: 重新塑造的输入和标签
"""
img_size, num_ch, num_class = \
            int(np.sqrt(x.shape[-1])), 1, len(np.unique(np.argmax(y, 1)))
dataset = x.reshape((-1, img_size, img_size, num_ch)).astype(np.float32)
labels = (np.arange(num_class) == y[:, None]).astype(np.float32)
return dataset, labels

def randomize(x, y):
    """随机化数据样本及其相应标签的顺序"""
    permutation = np.random.permutation(y.shape[0])
    shuffled_x = x[permutation, :, :, :]
    shuffled_y = y[permutation]
    return shuffled_x, shuffled_y

def get_next_batch(x, y, start, end):
    x_batch = x[start:end]
    y_batch = y[start:end]
    return x_batch, y_batch
```

加载数据并显示大小:

现在,可以在"训练"模式下使用定义的辅助函数。该函数可加载训练和验证图像以及相应的标签,还会显示数据集的大小:

```python
x_train, y_train, x_valid, y_valid = load_data(mode='train')
print("Size of:")
print("- Training-set:\t\t{}".format(len(y_train)))
print("- Validation-set:\t{}".format(len(y_valid)))
```

运行结果:

```
Extracting MNIST_data/train-images-idx3-ubyte.gz
Extracting MNIST_data/train-labels-idx1-ubyte.gz
Extracting MNIST_data/t10k-images-idx3-ubyte.gz
Extracting MNIST_data/t10k-labels-idx1-ubyte.gz
Size of:
- Training-set:         55000
- Validation-set:       5000
```

超参数:

```python
logs_path = "./logs"        #指向我们想要为Tensorboard保存日志的文件夹的路径
lr = 0.001                  #优化的初始学习率
```

```
epochs = 10                    #训练回合的总数
batch_size = 100               #训练批大小
display_freq = 100             #显示训练结果的频率
```

网络配置：

```
#第一个卷积层
filter_size1 = 5               #卷积核是 5×5 像素
num_filters1 = 16              #有 16 个这样的卷积核
stride1 = 1                    #滑动窗口的步幅

#第二个卷积层
filter_size2 = 5               #卷积核是 5×5 像素
num_filters2 = 32              #有 32 个这样的卷积核
stride2 = 1                    #滑动窗口的步幅

#全连接层
h1 = 128                       #完全连接层中的神经元数量
```

创建网络辅助函数。首先是用于创建新变量的辅助函数：

```
#权重和偏差包装器
def weight_variable(shape):
    """
    用适当的初始化创建一个权重变量
    :param name: 权重名称
    :param shape: 权重形状
    :return: 初始化后的权重变量
    """
    initer = tf.truncated_normal_initializer(stddev=0.01)
    return tf.get_variable('W',
                           dtype=tf.float32,
                           shape=shape,
                           initializer=initer)


def bias_variable(shape):
    """
    用适当的初始化器创建一个偏差变量
    :param name: 偏差变量名称
    :param shape: 偏差变量形状
    :return: 初始化后的偏差变量
    """
    initial = tf.constant(0., shape=shape, dtype=tf.float32)
    return tf.get_variable('b',
                           dtype=tf.float32,
                           initializer=initial)
```

用于创建新的卷积层的帮助函数:

```python
def conv_layer(x, filter_size, num_filters, stride, name):
    """
    创建一个2D卷积层
    :param x: 来自上一层的输入
    :param filter_size: 每个过滤器的大小
    :param num_filters: 过滤器的数量
    :param stride: 过滤器跨度
    :param name: 图层名称
    :return: 输出数组
    """
    with tf.variable_scope(name):
        num_in_channel = x.get_shape().as_list()[-1]
        shape = [filter_size, filter_size, num_in_channel, num_filters]
        W = weight_variable(shape=shape)
        tf.summary.histogram('weight', W)
        b = bias_variable(shape=[num_filters])
        tf.summary.histogram('bias', b)
        layer = tf.nn.conv2d(x, W,
                             strides=[1, stride, stride, 1],
                             padding="SAME")
        layer += b
        return tf.nn.relu(layer)
```

用于创建新的最大池图层的帮助函数:

```python
def max_pool(x, ksize, stride, name):
    """
    创建一个最大池化层
    :param x: 最大池化层的输入
    :param ksize: 最大池化层过滤器的大小
    :param stride: 最大池卷积核的步幅
    :param name: 图层名称
    :return: 输出数组
    """
    return tf.nn.max_pool(x,
                          ksize=[1, ksize, ksize, 1],
                          strides=[1, stride, stride, 1],
                          padding="SAME",
                          name=name)
```

用于展开图层的辅助函数:

```python
def flatten_layer(layer):
    """
    展平卷积层的输出,使其馈送到全连接层
```

```
:param layer: 输入数组
:return: 展平后的数组
"""
with tf.variable_scope('Flatten_layer'):
    layer_shape = layer.get_shape()
    num_features = layer_shape[1:4].num_elements()
    layer_flat = tf.reshape(layer, [-1, num_features])
    return layer_flat
```

用于创建新的全连接层的辅助函数：

```
def fc_layer(x, num_units, name, use_relu=True):
    """
    创建一个全连接层
    :param x: 来自上一层的输入
    :param num_units: 全连接层中隐藏单元的数量
    :param name: 图层名称
    :param use_relu: 用来决定是否添加 ReLU 非线性到图层的布尔值
    :return: 输出数组
    """
    with tf.variable_scope(name):
        in_dim = x.get_shape()[1]
        W = weight_variable(shape=[in_dim, num_units])
        tf.summary.histogram('weight', W)
        b = bias_variable(shape=[num_units])
        tf.summary.histogram('bias', b)
        layer = tf.matmul(x, W)
        layer += b
        if use_relu:
            layer = tf.nn.relu(layer)
        return layer
```

接下来创建 CNN 网络图。

输入（x）和相应标签（y）的占位符：

```
#Scope 可以帮助您在使用 TensorBoard 时分解您的模型
with tf.name_scope('Input'):
    x = tf.placeholder(tf.float32, shape=[None, img_h, img_w, n_channels], name='X')
    y = tf.placeholder(tf.float32, shape=[None, n_classes], name='Y')
```

创建网络层：

```
conv1 = conv_layer(x, filter_size1, num_filters1, stride1, name='conv1')
pool1 = max_pool(conv1, ksize=2, stride=2, name='pool1')
conv2 = conv_layer(pool1, filter_size2, num_filters2, stride2, name='conv2')
pool2 = max_pool(conv2, ksize=2, stride=2, name='pool2')
layer_flat = flatten_layer(pool2)
```

```
fc1 = fc_layer(layer_flat, h1, 'FC1', use_relu=True)
output_logits = fc_layer(fc1, n_classes, 'OUT', use_relu=False)
```

定义损失函数、优化器、准确度和预测类：

```
with tf.variable_scope('Train'):
    with tf.variable_scope('Loss'):
        loss = \
            tf.reduce_mean(tf.nn.softmax_cross_entropy_with_logits(labels=y, \
            logits=output_logits), name='loss')
    tf.summary.scalar('loss', loss)
    with tf.variable_scope('Optimizer'):
        optimizer = \
            tf.train.AdamOptimizer(learning_rate=lr, name='Adam-op').minimize(loss)
    with tf.variable_scope('Accuracy'):
        correct_prediction = \
            tf.equal(tf.argmax(output_logits, 1), tf.argmax(y, 1), name='correct_pred')
        accuracy = \
            tf.reduce_mean(tf.cast(correct_prediction, tf.float32), name='accuracy')
    tf.summary.scalar('accuracy', accuracy)
    with tf.variable_scope('Prediction'):
        cls_prediction = tf.argmax(output_logits, axis=1, name='predictions')
```

初始化所有变量并合并摘要：

```
#初始化变量
init = tf.global_variables_initializer()
#合并所有摘要
merged = tf.summary.merge_all()
```

训练：

```
sess = tf.InteractiveSession()
sess.run(init)
global_step = 0
summary_writer = tf.summary.FileWriter(logs_path, sess.graph)
#每个回合的训练迭代次数
num_tr_iter = int(len(y_train) / batch_size)
for epoch in range(epochs):
    print('Training epoch: {}'.format(epoch + 1))
    x_train, y_train = randomize(x_train, y_train)
    for iteration in range(num_tr_iter):
        global_step += 1
        start = iteration * batch_size
        end = (iteration + 1) * batch_size
        x_batch, y_batch = get_next_batch(x_train, y_train, start, end)
```

```python
        #运行优化运算(反向传播)
        feed_dict_batch = {x: x_batch, y: y_batch}
        sess.run(optimizer, feed_dict=feed_dict_batch)

        if iteration % display_freq == 0:
            #计算并显示批损失和准确度
            loss_batch, acc_batch, summary_tr = \
                sess.run([loss, accuracy, merged], \
                                feed_dict=feed_dict_batch)
            summary_writer.add_summary(summary_tr, global_step)

            print("iter {0:3d}:\t Loss={1:.2f},\tTraining Accuracy={2:.01%}".
                format(iteration, loss_batch, acc_batch))

    #在每个回合后运行验证
    feed_dict_valid = {x: x_valid, y: y_valid}
    loss_valid, acc_valid = sess.run([loss, accuracy], feed_dict=feed_dict_valid)
    print('---------------------------------------------------------')
    print("Epoch: {0}, validation loss: {1:.2f}, validation accuracy: {2:.01%}".
        format(epoch + 1, loss_valid, acc_valid))
    print('---------------------------------------------------------')
```

测试:

```python
#完成训练后测试网络
x_test, y_test = load_data(mode='test')
feed_dict_test = {x: x_test, y: y_test}
loss_test, acc_test = sess.run([loss, accuracy], feed_dict=feed_dict_test)
print('---------------------------------------------------------')
print("Test loss: {0:.2f}, test accuracy: {1:.01%}".format(loss_test, acc_test))
print('---------------------------------------------------------')

#绘制一些正确和错误分类的例子
cls_pred = sess.run(cls_prediction, feed_dict=feed_dict_test)
cls_true = np.argmax(y_test, axis=1)
plot_images(x_test, cls_true, cls_pred, title='Correct Examples')
plot_example_errors(x_test, cls_true, cls_pred, title='Misclassified Examples')
plt.show()
```

TensorFlow 读入一个图像列表:

```python
import tensorflow as tf

filenames = ['/home/aaa/firefox.jpg']
filename_queue = tf.train.string_input_producer(filenames)

reader = tf.WholeFileReader()
```

```
key, value = reader.read(filename_queue)

images = tf.image.decode_jpeg(value, channels=3)
```

TensorFlow 提供更高级别的 Estimator API，其中包含用于训练和预测数据的预建模型。

TensorFlow 的官方模型（https://github.com/tensorflow/models/tree/master/official）是使用 TensorFlow 高级 API 的示例模型的集合。

训练并保存模型：

```
$ python3 mnist.py --export_dir /home/ai/mnist_saved_model
```

或者使用 nohup 脱离控制台运行：

```
$ nohup python3 mnist.py --export_dir /home/ai/mnist_saved_model &
```

显示模型文件：

```
$ saved_model_cli show --dir /home/ai/mnist_saved_model/1530443317 --all
```

使用 TensorFlow 中的方法加载图像：

```
from tensorflow.python.lib.io import file_io

d = np.load(file_io.FileIO("F:/models-master/official/mnist/examples.npy", mode='rb'))
plt.imshow(d[1], cmap=plt.cm.gray)
plt.show()
```

运行模型文件：

```
$ saved_model_cli run --dir /home/ai/mnist_saved_model/1530443317 --tag_set serve --signature_def classify --inputs image=examples.npy
```

可能的输出结果如下：

```
Result for output key classes:
[5 3]
Result for output key probabilities:
[[ 1.53558474e-07   1.95694142e-13   1.31193523e-09   5.47467265e-03
   5.85711526e-22   9.94520664e-01   3.48423509e-06   2.65365645e-17
   9.78631419e-07   3.15522470e-08]
 [ 1.22413359e-04   5.87615965e-08   1.72251271e-06   9.39960718e-01
   3.30306928e-11   2.87386645e-02   2.82353517e-02   8.21146413e-18
   2.52568233e-03   4.15460236e-04]]
```

3.5.5　一维卷积

要手动计算一维卷积，可以在输入上滑动内核，计算逐元素乘法并对它们求和。

最简单的方法是 padding=0, stride=1。

因此，如果输入为[1,0,2,3,0,1,1]，并且卷积核为[2,1,3]，则卷积的结果是[8,11,7,9,4]，这是按以下方式计算的：

$$8 = 1 \times 2 + 0 \times 1 + 2 \times 3$$
$$11 = 0 \times 2 + 2 \times 1 + 3 \times 3$$
$$7 = 2 \times 2 + 3 \times 1 + 0 \times 3$$
$$9 = 3 \times 2 + 0 \times 1 + 1 \times 3$$
$$4 = 0 \times 2 + 1 \times 1 + 1 \times 3$$

TensorFlow 的 conv1d 函数分批计算卷积，所以为了在 TensorFlow 中执行此运算，需要以正确的格式提供数据。

处理批次的操作可能假设 TensorFlow 的第一个维度是批量维度。这里设置批大小为 1。计算一维卷积代码如下：

```
import tensorflow as tf
i = tf.constant([1, 0, 2, 3, 0, 1, 1], dtype=tf.float32, name='i')
k = tf.constant([2, 1, 3], dtype=tf.float32, name='k')

print (i, '\n', k, '\n')

data   = tf.reshape(i, [1, int(i.shape[0]), 1], name='data')
kernel = tf.reshape(k, [int(k.shape[0]), 1, 1], name='kernel')

print (data, '\n', kernel, '\n')

stride=1;
res = tf.squeeze(tf.nn.conv1d(data, kernel, stride, 'VALID'))
with tf.Session() as sess:
    print (sess.run(res))
```

这将返回先前计算的答案[8，11，7，9，4]。

用一些常量围绕输入矩阵称为填充（Padding）。在大多数情况下，此常量值为 0，这就是为什么大多数人将其命名为零填充。TensorFlow 支持 VALID 和 SAME 零填充，对于任意填充，则需要使用函数 tf.pad()。VALID 填充意味着根本没有填充。让我们在同一个例子中用 padding = 1 来计算卷积（请注意，对于我们的内核，这是 SAME 填充）。为此，我们只需在数组的开头和结尾添加 1 个 0：input = [0,1,0,2,3,0,1,1,0]。

注意，在这里不需要重新计算所有元素，即除了第一个和最后一个元素之外，所有元素保持不变：

$$1 = 0 \times 2 + 1 \times 1 + 0 \times 3$$

$3 = 1 \times 2 + 1 \times 1 + 0 \times 3$

因此，结果是[1, 8, 11, 7, 9, 4, 3] 与 TensorFlow 计算的结果相同：

```
res = tf.squeeze(tf.nn.conv1d(data, kernel, 1, 'SAME'))
with tf.Session() as sess:
    print sess.run(res)
```

使用步长的卷积时，步长（Stride）允许用户在滑动时跳过元素。在之前的所有示例中，我们滑动了 1 个元素，现在可以一次滑动 s 个元素。

因此，如果使用前面的示例 padding = 1 并将 stride 更改为 2，则只需获取前一个结果 [1,8,11,7,9,4,3]并每隔一个元素留下一个值，这将得到结果：[1,11,9,3]。可以通过以下方式在 TensorFlow 中执行此操作：

```
res = tf.squeeze(tf.nn.conv1d(data, kernel, 2, 'SAME'))
with tf.Session() as sess:
    print sess.run(res)
```

3.5.6 二维卷积

tf.nn.conv2d 函数实现卷积。在最基本的示例中，没有填充，而且 stride = 1。让假设输入和内核为：

$$\text{input} = \begin{pmatrix} 4 & 3 & 1 & 0 \\ 2 & 1 & 0 & 1 \\ 1 & 2 & 4 & 1 \\ 3 & 1 & 0 & 2 \end{pmatrix} \quad \text{kernel} = \begin{pmatrix} 1 & 0 & 1 \\ 2 & 1 & 0 \\ 0 & 0 & 1 \end{pmatrix}$$

应用卷积运算后，将收到以下输出：

$$\begin{pmatrix} 14 & 6 \\ 6 & 12 \end{pmatrix}$$

这个输出是通过以下方式计算的：

$14 = 4 \times 1 + 3 \times 0 + 1 \times 1 + 2 \times 2 + 1 \times 1 + 0 \times 0 + 1 \times 0 + 2 \times 0 + 4 \times 1$

$6 = 3 \times 1 + 1 \times 0 + 0 \times 1 + 1 \times 2 + 0 \times 1 + 1 \times 0 + 2 \times 0 + 4 \times 0 + 1 \times 1$

$6 = 2 \times 1 + 1 \times 0 + 0 \times 1 + 1 \times 2 + 2 \times 1 + 4 \times 0 + 3 \times 0 + 1 \times 0 + 0 \times 1$

$12 = 1 \times 1 + 0 \times 0 + 1 \times 1 + 2 \times 2 + 4 \times 1 + 1 \times 0 + 1 \times 0 + 0 \times 0 + 2 \times 1$

TensorFlow 的 conv2d 函数分批计算卷积并使用稍微不同的格式。对于输入，格式是[batch, in_height, in_width, in_channels]。对于内核，格式是[filter_height, filter_width, in_channels, out_channels]。所以需要以正确的格式提供数据：

```
import tensorflow as tf
k = tf.constant([
    [1, 0, 1],
    [2, 1, 0],
    [0, 0, 1]
], dtype=tf.float32, name='k')
i = tf.constant([
    [4, 3, 1, 0],
    [2, 1, 0, 1],
    [1, 2, 4, 1],
    [3, 1, 0, 2]
], dtype=tf.float32, name='i')
kernel = tf.reshape(k, [3, 3, 1, 1], name='kernel')
image  = tf.reshape(i, [1, 4, 4, 1], name='image')
```

然后通过以下方式计算卷积:

```
res = tf.squeeze(tf.nn.conv2d(image, kernel, [1, 1, 1, 1], "VALID"))
#VALID 表示没有填充
with tf.Session() as sess:
    print(sess.run(res))
```

输出结果将等同于我们手工计算的那个矩阵。

用一些常量围绕输入矩阵。在大多数情况下,常量为零,这就是人们称之为零填充的原因。因此,如果想在原始输入中使用 1 的填充(检查第一个示例,padding = 0,strides = 1),矩阵将如下所示:

$$\begin{pmatrix} 0 & 0 & 0 & 0 & 0 & 0 \\ 0 & 4 & 3 & 1 & 0 & 0 \\ 0 & 2 & 1 & 0 & 1 & 0 \\ 0 & 1 & 2 & 4 & 1 & 0 \\ 0 & 3 & 1 & 0 & 2 & 0 \\ 0 & 0 & 0 & 0 & 0 & 0 \end{pmatrix}$$

要计算卷积的值,请执行相同的滑动。请注意,在我们的情况下,中间的许多值不需要重新计算,因为它们与前面的示例相同。也不会在此显示所有计算,因为这个想法很简单。结果是:

$$\begin{pmatrix} 5 & 11 & 8 & 2 \\ 7 & 14 & 6 & 2 \\ 3 & 6 & 12 & 9 \\ 5 & 12 & 5 & 6 \end{pmatrix}$$

这里:

$$5 = 0\times1+0\times0+0\times1+0\times2+4\times1+3\times0+0\times0+0\times1+1\times1$$

......

$$6 = 4\times1+1\times0+0\times1+0\times2+2\times1+0\times0+0\times0+0\times0+0\times1$$

TensorFlow 不支持 conv2d 函数中的任意填充。因此，如果需要一些不支持的填充，请使用函数 tf.pad()。幸运的是，对于我们的输入，SAME 填充将等于 padding = 1。因此几乎不需要对前面的示例做任何改变：

```
res = tf.squeeze(tf.nn.conv2d(image, kernel, [1, 1, 1, 1], "SAME"))
#'SAME'确保我们的输出与输入具有相同的大小并使用适当的填充。
#在我们的例子中填充值是1。
with tf.Session() as sess:
    print sess.run(res)
```

可以验证答案是否与手动计算的答案相同。

图像处理中的卷积核参数在使用训练的方法调整以前，往往设置成一些固定的值。例如锐化参数：

```
sharpenMatrix = [0.0, -0.2, 0.0, -0.2, 1.8, -0.2, 0.0, -0.2, 0.0]
#使用初始化器(initializer)初始化卷积核
init = tf.constant_initializer(sharpenMatrix)
W= tf.get_variable('W', shape=[3, 3], initializer=init)
```

完整的代码如下：

```
sharpenMatrix = [0.0, -0.2, 0.0, -0.2, 1.8, -0.2, 0.0, -0.2, 0.0]
init = tf.constant_initializer(sharpenMatrix)
W= tf.get_variable('W', shape=[3, 3], initializer=init)

#添加一个操作来初始化全局变量
init_op = tf.global_variables_initializer()

#在会话中运行计算图
with tf.Session() as sess:
    #运行变量初始化器操作
    sess.run(init_op)
    #现在来评估变量的值
    print(sess.run(W))
```

3.5.7 膨胀卷积

膨胀卷积（Dilated Convolution）是针对图像语义分割问题中下采样会降低图像分辨率、丢失信息而提出的一种卷积思路。

完全卷积表明神经网络只由卷积层组成，没有任何完全连接的层或通常在网络末端找到的 MLP（Multi Layer Perceptron，多层感知器）。具有完全连接层的 CNN 就像完全卷积一样可以端到端地学习。和具有完全连接层的 CNN 的主要区别在于完全卷积网是在任何地方都是学习过滤器。甚至网络末端的决策层都是过滤器。

完全卷积网试图学习表示并根据局部空间输入作出决策。附加完全连接的层使得网络能够使用全局信息来学习某些东西，输入的空间布局消失于其中，当不需要用到输入的空间布局时，可以使用附加的完全连接层。

增加原本紧密贴着的卷积核元素之间的距离，但卷积核需要计算的点不变，也就是多余出来的位置全填 0。

卷积核的感知区域变大了，又由于卷积核中的有效计算点不变，所以计算量不变。每层的特征图的尺寸都不变，所以图像信息都保存了下来。

D 扩张的 $K×K$ 卷积在进行通常的卷积之前扩大卷积核。扩大卷积核意味着扩大其尺寸，用零填充空位置。实际上，并没有创建扩展的卷积核。相反，卷积核元素（权重）与输入矩阵中的远（不相邻）元素匹配。距离由扩张系数 D 确定。图 3-14 显示了卷积核元素如何与 D 扩张的 3×3 卷积中的输入元素匹配（当卷积核的中心与输入矩阵的中心对齐时）。请注意，对于 $D = 1$，将获得标准卷积。

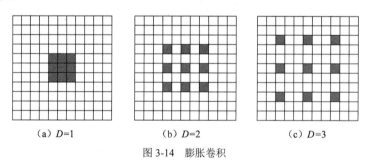

（a）D=1　　　（b）D=2　　　（c）D=3

图 3-14　膨胀卷积

在 D 膨胀卷积中，通常步幅是 1，但也可以使用其他步幅。

3.5.8　TensorFlow 实现简单的语音识别

使用 TensorFlow 识别如下 10 个英文单词：yes、no、up、down、left、right、on、off、stop、go。

首先使用 git 得到 TensorFlow 的源代码，然后在 TensorFlow 源码树下运行训练脚本：

```
#python3 tensorflow/examples/speech_commands/train.py
```

此时，系统会自动开始下载语音指令数据集。这个数据集含有超过 105000 个的 WAVE 音频文件，每个文件包含由人说的 30 个不同的单词。

这个存档文件超过 2GB，因此下载可能需要一段时间，但应该能看到进度日志，一旦下载完成，就不用再次执行此步骤了。程序会把这个数据文件下载到/tmp 临时目录下。

下载完成后，就将看到如下所示的训练日志记录信息：

```
I0730 16:53:44.766740    55030 train.py:176] Training from step: 1
I0730 16:53:47.289078    55030 train.py:217] Step #1: rate 0.001000, accuracy 7.0%, cross entropy 2.611571
```

这表明初始化过程已经完成，训练循环已经开始。这时将看到它为每个训练步骤输出的信息。如下是对日志记录信息的详细解释：

- Step #1 表明我们正处于训练循环的第一步。在这个例子中，总共将有 18000 个步骤，因此可以查看步骤编号，以了解训练距离完成还有多远。
- 速率 0.001000 是控制网络权重更新速度的学习速率。早期这是一个相对较高的数字（0.001），但对于后来的训练周期，它将减少到 1/10，即 0.0001。
- 准确度 7.0% 表示在此训练步骤中有多少类正确预测出来。这个值经常会波动很多，但随着训练的进行，平均值会增加。模型输出一个数字数组，每个标签一个，每个数字是输入的预测可能性。通过选择具有最高分数的条目来挑选预测标签。分数始终在 0 和 1 之间，较高的值表示对结果更有信心。
- 交叉熵 2.611571 是用来指导训练过程的损失函数的结果。这是通过将当前训练运行的得分矢量与正确标签进行比较而获得的得分，并且这个值应该在训练期间呈下降趋势。

经过 100 步后，应该看到如下信息：

```
I0730 16:54:41.813438    55030 train.py:252] Saving to "/tmp/speech_commands_train/conv.ckpt-100"
```

这样可以将当前训练的权重保存到检查点文件中。如果训练脚本被中断，可以查找上次保存的检查点，然后使用--start_checkpoint=/tmp/speech_commands_train/conv.ckpt-100 作为命令行参数从该点开始重新启动该脚本。

经过 400 步后，将记录混淆矩阵信息：

```
I0730 16:57:38.073667    55030 train.py:243] Confusion Matrix:
 [[258    0    0    0    0    0    0    0    0    0    0    0]
  [  7    6   26   94    7   49    1   15   40    2    0   11]
  [ 10    1  107   80   13   22    0   13   10    1    0    4]
  [  1    3   16  163    6   48    0    5   10    1    0   17]
  [ 15    1   17  114   55   13    0    9   22    5    0    9]
  [  1    1    6   97    3   87    1   12   46    0    0   10]
```

```
[  8    6   86   84   13   24   1    9    9    1    0    6  ]
[  9    3   32  112    9   26   1   36   19    0    0    9  ]
[  8    2   12   94    9   52   0    6   72    0    0    2  ]
[ 16    1   39   74   29   42   0    6   37    9    0    3  ]
[ 15    6   17   71   50   37   0    6   32    2    1    9  ]
[ 11    1    6  151    5   42   0    8   16    0    0   20 ]]
```

要理解混淆矩阵的含义，首先需要知道正在使用的标签，在这种情况下是"silence""unknown""yes""no""up""down""left""right""on""off""stop"和"go"。每列代表的一组预测为每个标签的样本，因此第一列代表预测为静音的所有片段，第二列代表所有预测为未知单词的片段，第三列代表"yes"，依此类推。

每行用正确的、真实标签表示音频剪辑。第一行是所有静音剪辑，第二行是未知单词，第三行代表"是"，等等。

脚本训练完 18000 步之后，会显示一份最终的混淆矩阵和一个根据测试集得出的准确率得分。如果按照默认设置进行训练，准确率应该在 80%～90%。

训练完成后，可以运行下面命令行，导出这个语音识别模型：

```
python3 tensorflow/examples/speech_commands/freeze.py \
--start_checkpoint=/tmp/speech_commands_train/conv.ckpt-18000 \
--output_file=/home/aaa/speech_commands/my_frozen_graph.pb
```

然后可以用 label_wav.py 脚本，让这个固定的模型识别音频试试：

```
python3 tensorflow/examples/speech_commands/label_wav.py \
--graph=/home/aaa/speech_commands/my_frozen_graph.pb \
--labels=/tmp/speech_commands_train/conv_labels.txt \
--wav=/tmp/speech_dataset/left/a5d485dc_nohash_0.wav
```

这个命令应该会输出 3 个标签的得分：

```
left (score = 0.79622)
right (score = 0.09350)
_unknown_ (score = 0.07849)
```

希望"left"是最高分，因为这是正确的标签，但由于训练是随机的，可以尝试使用同一文件夹中的一些其他.wav 文件来查看识别效果。标签得分在 0～1，较高的值意味着模型对其预测更有信心。

除了返回识别单词的回归值。还可以把识别出来的每个单词发音加注音标。

3.5.9　NumPy 提取语音识别特征

可以用 NumPy 提取语音识别特征，然后再交给 TensorFlow 处理。

这里使用来自开放式语音库（http://www.voiptroubleshooter.com/open_speech/american.html）的一个名为"OSR_us_000_0010_8k.wav"的 16 位 PCM 编码的.wav 文件，其采样频率为 8000 Hz。脉冲编码调制（PCM）是模拟信号的数字表示，PCM 以规则的间隔对模拟信号的幅度进行采样。

.wav 文件是一个干净的语音信号，包括一个单独的声音，说出一些句子，其间有一些暂停。为简单起见，使用信号的前 3.5s，大致对应于.wav 文件中的第一个句子。实现代码如下：

```
import numpy
import scipy.io.wavfile
from scipy.fftpack import dct

#假定文件位于同一目录中
sample_rate, signal = scipy.io.wavfile.read('OSR_us_000_0010_8k.wav')
signal = signal[0:int(3.5 * sample_rate)]  #保留前 3.5 秒
```

第一步是对信号应用预加重滤波器以放大高频。预加重滤波器在以下几方面很有用：①平衡频谱，因为高频通常与较低频率相比具有较小的幅度；②避免在傅里叶变换运算期间的数值问题；③可以改善信噪比（SNR）。

可以使用以下等式中的一阶滤波器将预加重滤波器应用于信号 x：

$$y(t)=x(t)-\alpha x(t-1)$$

这可以使用以下行轻松实现，其中滤波器系数 α 的典型值为 0.95 或 0.97：

```
pre_emphasis = 0.97
emphasized_signal = numpy.append(signal[0], signal[1:] - pre_emphasis * signal[:-1])
```

在预加重之后，需要将信号分成短时帧。这一步骤的基本原理是信号中的频率随时间变化，因此在大多数情况下，对整个信号进行傅里叶变换是没有意义的，因为会随着时间的推移丢失信号的频率轮廓。为了避免这种情况，可以安全地假设信号中的频率在很短的时间内是静止的。因此，通过在该短时帧上进行傅里叶变换，可以通过连接相邻帧来获得信号频率轮廓的良好近似。

语音处理中的典型帧大小范围为 20~40ms，连续帧之间具有 40%~60%的重叠。帧大小的流行设置为 25ms 和 10 ms 步幅（15 ms 重叠）。

```
frame_size = 0.025        #帧大小
frame_stride = 0.01       #步幅
#从秒转换为样本
frame_length, frame_step = frame_size * sample_rate, frame_stride * sample_rate
signal_length = len(emphasized_signal)
frame_length = int(round(frame_length))
frame_step = int(round(frame_step))
num_frames = int(numpy.ceil(float(numpy.abs(signal_length - frame_length)) /
```

```
frame_step))    #确保至少有一帧

    pad_signal_length = num_frames * frame_step + frame_length
    z = numpy.zeros((pad_signal_length - signal_length))
    pad_signal = numpy.append(emphasized_signal, z)   #填充信号以确保所有帧具有相同数量的样本而
                                                       #不截断原始信号中的任何样本

    indices = numpy.tile(numpy.arange(0, frame_length), (num_frames, 1)) + numpy.tile
(numpy.arange(0, num_frames * frame_step, frame_step), (frame_length, 1)).T
    frames = pad_signal[indices.astype(numpy.int32, copy=False)]
```

在将信号切换成帧之后，将诸如海明窗口的窗口函数应用于每帧。海明窗口具有以下形式：

$$w[n] = 0.54 - 0.46 \cos\left(\frac{2\pi n}{N-1}\right)$$

这里，$0 \leq n \leq N-1$，N 是窗口长度。

需要对帧应用窗口函数有几个原因，特别是为了抵消 FFT 数据无限的假设并减少频谱泄漏。

```
frames *= numpy.hamming(frame_length)
#frames *= 0.54 - 0.46 * numpy.cos((2 * numpy.pi * n) / (frame_length - 1))
                                                        #明确的实现  **
```

现在可以对每帧进行 N 点 FFT 计算频谱，也称为短时傅里叶变换（STFT），其中 N 通常为 256 或 512，NFFT = 512；然后使用以下等式计算功率谱（周期图）：

$$P = \frac{|FFT(x_i)|^2}{N}$$

其中，x_i 是信号 x 的第 i 帧。可以通过以下几行代码来实现：

```
mag_frames = numpy.absolute(numpy.fft.rfft(frames, NFFT))   #FFT 的大小
pow_frames = ((1.0 / NFFT) * ((mag_frames) ** 2))            #功率谱
```

计算滤波器组（Filter Banks）的最后一步是应用三角滤波器，通常为 40 个滤波器，在梅尔刻度上以 nfilt = 40 来提取功率谱以提取频带。梅尔刻度旨在通过在较低频率处更具辨别力并在较高频率处较少辨别力来模仿非线性人耳对声音的感知。提取 fbank 特征的代码如下：

```
low_freq_mel = 0
high_freq_mel = (2595 * numpy.log10(1 + (sample_rate / 2) / 700))   #将频率转换为 Mel
mel_points = numpy.linspace(low_freq_mel, high_freq_mel, nfilt + 2)
                                                       #梅尔刻度的间距相等
hz_points = (700 * (10**(mel_points / 2595) - 1))       #将 Mel 转换为频率
bin = numpy.floor((NFFT + 1) * hz_points / sample_rate)

fbank = numpy.zeros((nfilt, int(numpy.floor(NFFT / 2 + 1))))
for m in range(1, nfilt + 1):
```

```
    f_m_minus = int(bin[m - 1])        #左
    f_m = int(bin[m])                  #中
    f_m_plus = int(bin[m + 1])         #右

    for k in range(f_m_minus, f_m):
        fbank[m - 1, k] = (k - bin[m - 1]) / (bin[m] - bin[m - 1])
    for k in range(f_m, f_m_plus):
        fbank[m - 1, k] = (bin[m + 1] - k) / (bin[m + 1] - bin[m])
filter_banks = numpy.dot(pow_frames, fbank.T)
filter_banks = numpy.where(filter_banks == 0, numpy.finfo(float).eps, filter_banks)
#数值稳定性
filter_banks = 20 * numpy.log10(filter_banks)   #dB
```

3.5.10 Numba

Numba 是能够感知 NumPy 的 Python 优化编译器。它使用 LLVM 编译器项目从 Python 语法生成机器代码。

Numba 可以编译大量以数字为重点的 Python 代码，包括许多 NumPy 函数。此外，Numba 支持自动循环并行化，生成 GPU 加速代码，以及创建 ufunc 和 C 回调。

安装 Numba 并获取更新的最简单方法是使用 Anaconda 分发：

```
$ conda install numba
```

还可以使用 pip 命令安装 Numba：

```
$ pip install numba
```

Numba 适用于看起来像这样的代码：

```
from numba import jit
import numpy as np

x = np.arange(100).reshape(10, 10)

@jit(nopython=True)  #设置 nopython 模式以获得最佳性能，相当于@njit
def go_fast(a):       #第一次调用时，函数被编译为机器代码
    trace = 0
    for i in range(a.shape[0]):       #Numba 喜欢循环
        trace += np.tanh(a[i, i])     #Numba 喜欢 NumPy 的函数
    return a + trace                  #Numba 喜欢 NumPy 广播

print(go_fast(x))
```

3.6 端到端深度学习

可以使用交叉熵目标训练前馈 DNN（卷积神经网络）声学模型。DNN 包括简单的 DNN、TDNN（时延神经网络）和 CNN。

CNN 在声学模型上的应用源于近些年来 CNN 在图像处理上压倒性的成功。最初应用于声学模型的 CNN 只有一层或者两层，用于特征的提取，在此之上再加上 LSTM 和标准的前向网络，代表的例子是 CNN-LSTM-DNN（CLDNN）。由于极其深的 CNN（诸如 VGGnet 和 Resnet）在图像识别上的成功，这些模型也被引入到声学模型，IBM 和微软的学者还在此之上提出了一系列其他变形。这些 CNN 模型在 SwitchBoard 这个标准语音识别任务上不断刷新最低错误率的记录。

3.7 Dropout 解决过度拟合问题

过度拟合是指模型过分地拟合训练样本，但对测试样本预测准确率不高。过度拟合导致模型泛化能力差。Dropout 是一种通过暂时不使用的一些节点来解决神经网络过度拟合问题的方法。

将 Dropout 视为一种集成学习形式是有帮助的。在集成学习中，我们采用了一些"较弱"的分类器，分别训练它们，然后在测试时我们通过平均所有集合成员的响应来使用它们。由于每个分类器都经过单独训练，因此它学会了数据的不同"方面"，并且它们的错误也不同。将它们组合起来有助于产生更强的分级器，不易过度拟合。随机森林或梯度增强树（Gradient-Boosted Tree，GBT）是典型的集成分类器。

集成学习的一个变体是装袋法，其中集成分类器中的每个成员用输入数据的不同子样本训练，因此仅学习了整个可能的输入特征空间的子集。

Dropout 可以被视为装袋法的极端版本。在小批量的每个训练步骤中，Dropout 程序创建不同的网络（通过随机移除一些单元），其像往常一样使用反向传播进行训练。从概念上讲，整个过程类似于使用许多不同网络的集合（每步一个），每个网络用单个样本训练（即极端装袋）。

在测试阶段，使用整个网络（所有单位），但权重按比例缩小。在数学上，这近似于整体平均（使用几何平均值作为平均值）。

使用 tf.layers.dropout 实现 Dropout 的代码如下：

```
#导入 MNIST 数据
from tensorflow.examples.tutorials.mnist import input_data
```

```python
mnist = input_data.read_data_sets("/tmp/data/", one_hot=False)

import tensorflow as tf

#训练参数
learning_rate = 0.001
num_steps = 2000
batch_size = 128

#网络参数
num_input = 784     #MNIST 数据输入(img shape: 28×28)
num_classes = 10    #MNIST 类别总数（0-9 数字）
dropout = 0.25      #暂时不使用一个单位的可能性

#创建神经网络
def conv_net(x_dict, n_classes, dropout, reuse, is_training):
    #定义重用变量的范围
    with tf.variable_scope('ConvNet', reuse=reuse):
        #在多个输入的情况下，TF Estimator 输入是一个字典
        x = x_dict['images']

        #MNIST 数据输入是一个具有 784 个特征的一维向量(28×28 像素)
        #重塑以匹配图片格式[高度 x 宽度 x 通道数]
        #张量输入变为四维： [批大小，高度，宽度，通道数]
        x = tf.reshape(x, shape=[-1, 28, 28, 1])

        #卷积层有 32 个过滤器，卷积核大小为 5
        conv1 = tf.layers.conv2d(x, 32, 5, activation=tf.nn.relu)
        #最大池化（下采样），步长为 2，核大小为 2
        conv1 = tf.layers.max_pooling2d(conv1, 2, 2)

        #卷积层有 64 个过滤器，卷积核大小为 3
        conv2 = tf.layers.conv2d(conv1, 64, 3, activation=tf.nn.relu)
        #最大池化（下采样），步长为 2，核大小为 2
        conv2 = tf.layers.max_pooling2d(conv2, 2, 2)

        #为了全连接层，将数据展平为一维向量
        fc1 = tf.contrib.layers.flatten(conv2)

        #全连接图层（现在在 tf contrib 文件夹中）
        fc1 = tf.layers.dense(fc1, 1024)
        #应用 Dropout（如果 is_training 为 False，则不应用 dropout）
        fc1 = tf.layers.dropout(fc1, rate=dropout, training=is_training)

        #输出层，类预测
```

```python
        out = tf.layers.dense(fc1, n_classes)

    return out

#定义模型函数(遵循 TF Estimator 模板)
def model_fn(features, labels, mode):
    #构建神经网络
    #因为 Dropout 在训练和预测时有不同的行为,
    #所以需要创建两个仍然共享相同权重的不同计算图
    logits_train = conv_net(features, num_classes, dropout, reuse=False,
                            is_training=True)
    logits_test = conv_net(features, num_classes, dropout, reuse=True,
                           is_training=False)

    #预测
    pred_classes = tf.argmax(logits_test, axis=1)
    pred_probas = tf.nn.softmax(logits_test)

    #如果是预测模式,则提前返回
    if mode == tf.estimator.ModeKeys.PREDICT:
        return tf.estimator.EstimatorSpec(mode, predictions=pred_classes)

        #定义损失和优化器
    loss_op = tf.reduce_mean(tf.nn.sparse_softmax_cross_entropy_with_logits(
        logits=logits_train, labels=tf.cast(labels, dtype=tf.int32)))
    optimizer = tf.train.AdamOptimizer(learning_rate=learning_rate)
    train_op = optimizer.minimize(loss_op,
                                  global_step=tf.train.get_global_step())

    #评估模型的准确性
    acc_op = tf.metrics.accuracy(labels=labels, predictions=pred_classes)

    #Estimators 需要返回 EstimatorSpec,然后用它指定训练和评估等的不同操作
    estim_specs = tf.estimator.EstimatorSpec(
        mode=mode,
        predictions=pred_classes,
        loss=loss_op,
        train_op=train_op,
        eval_metric_ops={'accuracy': acc_op})

    return estim_specs

#构建 Estimator
model = tf.estimator.Estimator(model_fn)
```

```
#定义用于训练的输入函数
input_fn = tf.estimator.inputs.numpy_input_fn(
    x={'images': mnist.train.images}, y=mnist.train.labels,
    batch_size=batch_size, num_epochs=None, shuffle=True)
#Train the Model
model.train(input_fn, steps=num_steps)

#评估模型
#定义用于评估的输入函数
input_fn = tf.estimator.inputs.numpy_input_fn(
    x={'images': mnist.test.images}, y=mnist.test.labels,
    batch_size=batch_size, shuffle=False)
#使用 Estimator 的 evaluate 方法
e = model.evaluate(input_fn)

print("Testing Accuracy:", e['accuracy'])
```

3.8 NumPy 中的矩阵运算

矩阵的点乘就是矩阵中的各个对应元素相乘，使用函数 numpy.dot() 计算两个矩阵的点乘。示例代码如下：

```
>>> import numpy as np
>>> A = np.array([[1,2,3,4,5,6],
...               [1,2,3,4,5,6],
...               [1,2,3,4,5,6],
...               [1,2,3,4,5,6],
...               [1,2,3,4,5,6],
...               [1,2,3,4,5,6]])
>>>
>>>
>>> b = np.array([7,8,9,10,11,12]).reshape(6, 1)
>>>
>>> B = np.dot(A, b)
>>>
>>> B
array([[217],
       [217],
       [217],
       [217],
       [217],
       [217]])
```

矩阵的叉乘就是矩阵 *a* 的第 *i* 行和矩阵 *b* 的第 *j* 列的各个元素对应相乘然后求和作为输出矩阵第 *i* 行、第 *j* 列元素的值。函数 numpy.matmul()用于返回两个数组的矩阵叉乘乘积。

```
>>> h = [[1,2],[3,4]]
>>> i = [[5,6],[7,8]]
>>> np.matmul(h, i)     #结果矩阵第 1 行第 1 列元素的计算方法是：1×5+2×7 = 19
array([[19, 22],
       [43, 50]])
```

计算矩阵的秩：

```
>>> A = numpy.matrix([[1,3,7],[2,8,3],[7,8,1]])
>>> numpy.linalg.matrix_rank(A)
3
```

行数与列数相等的矩阵称之为方阵。矩阵的行列式是可以从方阵计算出的一个标量。可以使用 np.linalg.det()计算行列式。

```
>>> i = [[5,6],[7,8]]
>>> np.linalg.det(i)    #矩阵 i 的行列式计算方法是：5×8-7×6
-2.000000000000005
```

将矩阵的行列互换得到的新矩阵称为转置矩阵，转置矩阵的行列式不变。如果要将 1 维向量转换为 2 维数组然后转置这个数组，只需使用 np.newaxis 对 1 维向量进行切片。

```
>>> a = np.array([5,4])[np.newaxis]
>>> print(a)
[[5 4]]
>>> print(a.T)
[[5]
 [4]]
```

3.9 说话者识别

不同的说话者在一个通用声学簇内具有不同的子空间。从通用簇偏移的子空间描述了样本的方向向量，其取决于说话者的语言文本。为了获得相关的仅包含说话者的特征，将这些向量分析为特征因子。分析的因子特征称为身份向量（identity vectors，即 I-vectors）。I-vectors 在语音段中传达说话者特性。

可以使用 i-vectors 来识别说话者，例如，通过计算两段语音表示的两个向量之间的余弦距离来作为这两段语音是否来源于同一个说话者的衡量指标。

用于说话人验证的深度学习和三维 CNN（https://github.com/astorfi/3D-convolutional-speaker-recognition）使用三维 CNN 的实现用于说话人模型。这里使用 TF-Slim 简化构建、训练和评估神经网络：

```
#构建一个三维卷积层
net = slim.conv2d(inputs, 16, [3, 1, 5], stride=[1, 1, 1], scope='conv11')
#带参数的 ReLU 激活函数
net = PReLU(net, 'conv11_activation')
net = slim.conv2d(net, 16, [3, 9, 1], stride=[1, 2, 1], scope='conv12')
net = PReLU(net, 'conv12_activation')
#在输入上执行三维最大池化
net = tf.nn.max_pool3d(net, strides=[1, 1, 1, 2, 1], ksize=[1, 1, 1, 2, 1], padding='VALID', name='pool1')

############Conv-2 ##############
############Conv-1 ##############
net = slim.conv2d(net, 32, [3, 1, 4], stride=[1, 1, 1], scope='conv21')
net = PReLU(net, 'conv21_activation')
net = slim.conv2d(net, 32, [3, 8, 1], stride=[1, 2, 1], scope='conv22')
net = PReLU(net, 'conv22_activation')
net = tf.nn.max_pool3d(net, strides=[1, 1, 1, 2, 1], ksize=[1, 1, 1, 2, 1], padding='VALID', name='pool2')

############Conv-3 ##############
############Conv-1 ##############
net = slim.conv2d(net, 64, [3, 1, 3], stride=[1, 1, 1], scope='conv31')
net = PReLU(net, 'conv31_activation')
net = slim.conv2d(net, 64, [3, 7, 1], stride=[1, 1, 1], scope='conv32')
net = PReLU(net, 'conv32_activation')
#net = slim.max_pool2d(net, [1, 1], stride=[4, 1], scope='pool1')

############Conv-4 ##############
net = slim.conv2d(net, 128, [3, 1, 3], stride=[1, 1, 1], scope='conv41')
net = PReLU(net, 'conv41_activation')
net = slim.conv2d(net, 128, [3, 7, 1], stride=[1, 1, 1], scope='conv42')
net = PReLU(net, 'conv42_activation')
#net = slim.max_pool2d(net, [1, 1], stride=[4, 1], scope='pool1')

############Conv-5 ##############
net = slim.conv2d(net, 128, [4, 3, 3], stride=[1, 1, 1], normalizer_fn=None, scope='conv51')
net = PReLU(net, 'conv51_activation')
```

```
#最后一层是类的logits
logits = tf.contrib.layers.conv2d(net, num_classes, [1, 1, 1], activation_fn=None,
scope='fc')
```

上面的代码使用了函数 slim.conv2d()。然而，简单地通过将三维内核用作[k_x, k_y, k_z]和 stride=[a, b, c]，可以将这个函数转换为三维卷积操作。

3.10 联邦学习

联邦学习（FL）让许多参与的客户能够训练共享的 ML 模型，同时把数据保存在本地。联邦数据通常是非独立同分布的。

使用 FEMNIST 训练示例来介绍 TFF 的联邦学习 API 层。这里使用的 FEMNIST 数据集是：62 个不同的类（包括 10 个数字、26 个小写字母、26 个大写字母），图像是 28×28 像素（可选让图像全部为 128×128 像素），由 3500 个书写者所写。由于每个书写者都有一个独特的风格，这个数据集展示了联合数据集非独立同分布类型的预期行为。

安装 tensorflow_federated：

```
#pip install --quiet tensorflow_federated
#pip install --quiet tf-nightly
```

测试：

```
from __future__ import absolute_import, division, print_function

import collections

from six.moves import range
import numpy as np
import tensorflow as tf

import tensorflow_federated as tff

np.random.seed(0)

tf.compat.v1.enable_v2_behavior()

tff.federated_computation(lambda: 'Hello, World!')()
```

输出：

```
'Hello, World!'
```

加载输入数据：

```
#@test {"output": "ignore"}
emnist_train, emnist_test = tff.simulation.datasets.emnist.load_data()
```

load_data()返回的数据集是 tff.simulation.ClientData 的实例，这个接口允许枚举用户集，构造表示特定用户数据的 tf.data.Dataset，并查询个别元素的结构。以下是如何使用此界面探索数据集的内容。请记住，虽然此接口允许迭代客户端 ID，但这只是模拟数据的一个功能。很快就会看到，联合学习框架不使用客户端身份——它们的唯一目的是允许选择数据的子集进行模拟。

```
len(emnist_train.client_ids)
```

输出：

```
3383
emnist_train.output_types, emnist_train.output_shapes
```

输出：

```
(OrderedDict([(u'label', tf.int32), (u'pixels', tf.float32)]),
 OrderedDict([(u'label', TensorShape([])), (u'pixels', TensorShape([28, 28]))]))
```

执行如下代码：

```
example_dataset = emnist_train.create_tf_dataset_for_client(
    emnist_train.client_ids[0])

example_element = iter(example_dataset).next()

example_element['label'].numpy()

emnist_train.output_types, emnist_train.output_shapes
```

输出：

```
5
```

执行如下代码：

```
#@test {"output": "ignore"}
from matplotlib import pyplot as plt
plt.imshow(example_element['pixels'].numpy(), cmap='gray', aspect='equal')
plt.grid('off')
_ = plt.show()
```

输出：

由于数据已经是 tf.data.Dataset，因此可以使用数据集转换完成预处理。在这里，我们将 28×28 像素的图像压缩成有 784 个元素的数组，将各个示例混合，将它们组织成批，然后将像素和标签的特征重命名为 x 和 y，以便与 Keras 一起使用。我们还重复数据集以运行几个回合。代码如下：

```
NUM_EPOCHS = 10
BATCH_SIZE = 20
SHUFFLE_BUFFER = 500

def preprocess(dataset):

  def element_fn(element):
    return collections.OrderedDict([
        ('x', tf.reshape(element['pixels'], [-1])),
        ('y', tf.reshape(element['label'], [1])),
    ])

  return dataset.repeat(NUM_EPOCHS).map(element_fn).shuffle(
      SHUFFLE_BUFFER).batch(BATCH_SIZE)
```

让我们验证一下是否有效。

```
#@test {"output": "ignore"}
preprocessed_example_dataset = preprocess(example_dataset)

sample_batch = tf.nest.map_structure(
    lambda x: x.numpy(), iter(preprocessed_example_dataset).next())

sample_batch
```

输出：

```
OrderedDict([('x', array([[ 1.,  1.,  1., ...,  1.,  1.,  1.],
```

```
          [ 1.,  1.,  1., ...,  1.,  1.,  1.],
          [ 1.,  1.,  1., ...,  1.,  1.,  1.],
          ...,
          [ 1.,  1.,  1., ...,  1.,  1.,  1.],
          [ 1.,  1.,  1., ...,  1.,  1.,  1.],
          [ 1.,  1.,  1., ...,  1.,  1.,  1.]], dtype=float32)), ('y', array([[3],
          [0],
          [7],
          [0],
          [8],
          [2],
          [7],
          [7],
          [9],
          [0],
          [5],
          [3],
          [3],
          [7],
          [1],
          [2],
          [6],
          [5],
          [2],
          [0]], dtype=int32))])
```

我们几乎获得了由所有构建块来构建的联邦数据集。

在模拟中将联邦数据提供给 TFF 的方法之一就是 Python 列表，列表中的每个元素都包含单个用户的数据，无论是作为列表还是作为 tf.data.Dataset。由于我们已经有了一个提供后者的接口，那么我们可以使用它。

这是一个简单的辅助函数，它将构建一组给定用户的数据集列表，作为一轮训练或评估的输入。

```
def make_federated_data(client_data, client_ids):
    return [preprocess(client_data.create_tf_dataset_for_client(x))
            for x in client_ids]
```

现在，我们如何选择客户设备？

在典型的联邦训练场景中，我们正在处理的可能是非常大量的用户设备，其中只有一小部分可用于在给定时间点进行训练。例如，当客户端设备是仅在插入电源，关闭计量网络以及以其他方式空闲时参与训练的移动电话时就是这种情况。

当然，我们处于模拟环境中，所有数据都是本地可用的。通常情况下，在运行模拟时，我们只需对客户的随机子集进行抽样，以参与每轮训练，但每轮训练通常不同。

也就是说，在每轮中随机抽样的客户子集的系统中实现收敛可能需要一段时间。

这里要做的是对客户端集进行一次采样，并在多轮中重复使用相同的集合以加速收敛（故意过度拟合这些少数用户的数据）。读者可以修改这里的例子以模拟随机抽样。

```
#@test {"output": "ignore"}
NUM_CLIENTS = 3

sample_clients = emnist_train.client_ids[0:NUM_CLIENTS]

federated_train_data = make_federated_data(emnist_train, sample_clients)

len(federated_train_data), federated_train_data[0]
```

输出如下：

```
(3,
 <DatasetV1Adapter shapes: OrderedDict([(x, (None, 784)), (y, (None, 1))]), types: OrderedDict([(x, tf.float32), (y, tf.int32)])>)
```

使用 Keras 创建模型：

如果您使用 Keras，您可能已经拥有构建 Keras 模型的代码。下面一个简单模型的例子，足以满足我们的需求。

```
def create_compiled_keras_model():
  model = tf.keras.models.Sequential([
      tf.keras.layers.Dense(
          10, activation=tf.nn.softmax, kernel_initializer='zeros', input_shape=(784,))])

  model.compile(
      loss=tf.keras.losses.SparseCategoricalCrossentropy(),
      optimizer=tf.keras.optimizers.SGD(learning_rate=0.02),
      metrics=[tf.keras.metrics.SparseCategoricalAccuracy()])
  return model
```

为了使用任何带有 TFF 的模型，需要将其包装在 tff.learning.Model 接口的实例中，该接口提供公开方法以标记模型的正向传递，元数据属性等，类似于 Keras，但也引入了额外的元素，例如控制计算联邦度量的过程的方法。我们暂时不用担心这个问题；如果你有一个我们上面刚刚定义的编译过的 Keras 模型，可以通过调用 tff.learning.from_compiled_keras_model 让 TFF 为你包装它，将模型和样本数据批量作为参数传递，如下所示。

```
def model_fn():
    keras_model = create_compiled_keras_model()
    return tff.learning.from_compiled_keras_model(keras_model, sample_batch)
```

接下来在联邦数据上训练模型。

现在我们有一个包装为 tff.learning.Model 的模型用于 TFF，我们可以通过调用辅助函数 tff.learning.build_federated_averaging_process 让 TFF 构造一个联邦平均算法，如下所示。

请记住，参数需要是构造函数（例如上面的 model_fn），而不是已构造的实例，因此模型的构造可以在由 TFF 控制的上下文中发生。

```
#@test {"output": "ignore"}
iterative_process = tff.learning.build_federated_averaging_process(model_fn)
```

这里，TFF 构建了一对联邦计算并将它们打包成 tff.utils.IterativeProcess，其中这些计算可作为一对 initialize 和 next 属性使用。

让我们从初始化计算开始。与所有联邦计算的情况一样，可以将其视为一个函数。计算不带参数，并返回一个结果——服务器上联邦平均过程的状态表示。

让我们调用初始化计算来构造服务器状态。

```
state = iterative_process.initialize()
```

next 属性将服务器状态（包括模型参数）推送到客户端，对其本地数据进行设备上训练，收集和平均模型更新，并在服务器上生成新的更新模型。

让我们进行一轮训练，并将结果可视化。我们可以将上面已经生成的联邦数据用于用户样本。

```
#@test {"timeout": 600, "output": "ignore"}
state, metrics = iterative_process.next(state, federated_train_data)
print('round  1, metrics={}'.format(metrics))
```

输出结果如下：

round 1, metrics=<sparse_categorical_accuracy=0.142909,loss=3.14069>

再进行几轮训练。

```
#@test {"skip": true}
for round_num in range(2, 11):
    state, metrics = iterative_process.next(state, federated_train_data)
    print('round {:2d}, metrics={}'.format(round_num, metrics))
```

输出结果如下：

```
round  2, metrics=<sparse_categorical_accuracy=0.166909,loss=2.90004>
round  3, metrics=<sparse_categorical_accuracy=0.203273,loss=2.64551>
round  4, metrics=<sparse_categorical_accuracy=0.248364,loss=2.41201>
round  5, metrics=<sparse_categorical_accuracy=0.291636,loss=2.19657>
round  6, metrics=<sparse_categorical_accuracy=0.341818,loss=1.99344>
round  7, metrics=<sparse_categorical_accuracy=0.397455,loss=1.81096>
round  8, metrics=<sparse_categorical_accuracy=0.446182,loss=1.65356>
round  9, metrics=<sparse_categorical_accuracy=0.486182,loss=1.51823>
round 10, metrics=<sparse_categorical_accuracy=0.533455,loss=1.39974>
```

每轮联邦训练后训练损失减少，表明模型正在收敛。

要对联邦数据执行评估，您可以使用 tff.learning.build_federated_evaluation 函数构建另一个为此目的而设计的联邦计算，并将模型构造函数作为参数传递。请注意，与使用 MnistTrainableModel 的联邦平均不同，它足以传递 MnistModel。评估不执行梯度下降，并且不需要构建优化器。

3.11 本章小结

1943 年，Warren McCulloch 和 Walter Pitts 创建了一种称为阈值逻辑的基于数学和算法的神经网络计算模型。1975 年，Werbos 的反向传播算法通过使多层网络的训练可行和有效从而解决了 XOR 问题。反向传播是一种相对于神经网络中的权重来计算损失函数的梯度的方法。

前馈神经网络是一种最简单的神经网络，各神经元分层排列。每个神经元只与前一层的神经元相连。接收前一层的输出，并输出给下一层。各层间没有反馈。卷积神经网络是受语音信号处理中时延神经网络（TDNN）影响而发明的。

本章介绍了语音识别中感知声音可以用到的深度学习方法及其实现。

第 4 章 C#开发深度学习应用

首先介绍在 C#语言中通过 TensorFlow.NET 使用 TensorFlow。

4.1 使用 TensorFlow.NET

TensorFlow.NET（简称 TF.NET）为 TensorFlow 提供.NET 标准绑定。

NuGet 是一个 Visual Studio 扩展，它使得在 Visual Studio 中安装和更新第三方库和工具更容易。可以在命令行使用 NuGet 安装项目需要的包。

通过 NuGet 安装 TF.NET。

```
PM> Install-Package TensorFlow.NET
```

导入 TF.NET。

```
using Tensorflow;
```

添加两个常量：

```
//创建一个常量操作
var a = tf.constant(4.0f);
var b = tf.constant(5.0f);
var c = tf.add(a, b);

using (var sess = tf.Session())
{
    var o = sess.run(c);
}
```

馈送占位符:

```
//创建一个占位符操作
var a = tf.placeholder(tf.float32);
var b = tf.placeholder(tf.float32);
var c = tf.add(a, b);

using(var sess = tf.Session())
{
    var o = sess.run(c, new FeedItem(a, 3.0f), new FeedItem(b, 2.0f));
}
```

线性回归:

```
//设置一个固定的初始值来调试
var W = tf.Variable(-0.06f, name: "weight");
var b = tf.Variable(-0.73f, name: "bias");

//构建一个线性模型
var pred = tf.add(tf.multiply(X, W), b);

//均方误差
var cost = tf.reduce_sum(tf.pow(pred - Y, 2.0f)) / (2.0f * n_samples);

//梯度下降
var optimizer = tf.train.GradientDescentOptimizer(learning_rate).minimize(cost);

//初始化变量(即给这些变量分配默认值)
var init = tf.global_variables_initializer();

//开始训练
with(tf.Session(), sess =>
{
    //运行初始化程序
    sess.run(init);

    //使用所有的训练数据
    for (int epoch = 0; epoch < training_epochs; epoch++)
    {
        foreach (var (x, y) in zip<float>(train_X, train_Y))
            sess.run(optimizer, new FeedItem(X, x), new FeedItem(Y, y));

        //显示每个回合步骤的日志
        if ((epoch + 1) % display_step == 0)
        {
            var c = sess.run(cost, new FeedItem(X, train_X), new FeedItem(Y, train_Y));
```

```csharp
                Console.WriteLine($"Epoch: {epoch + 1} cost={c} " + $"W={sess.run(W)} b={sess.run(b)}");
            }

            Console.WriteLine("Optimization Finished!");
            var training_cost = sess.run(cost, new FeedItem(X, train_X), new FeedItem(Y, train_Y));
            Console.WriteLine($"Training cost={training_cost} W={sess.run(W)} b={sess.run(b)}");

            //测试例子
            var test_X = np.array(6.83f, 4.668f, 8.9f, 7.91f, 5.7f, 8.7f, 3.1f, 2.1f);
            var test_Y = np.array(1.84f, 2.273f, 3.2f, 2.831f, 2.92f, 3.24f, 1.35f, 1.03f);
            Console.WriteLine("Testing... (Mean square loss Comparison)");

            var testing_cost = sess.run(tf.reduce_sum(tf.pow(pred - Y, 2.0f)) / (2.0f * test_X.shape[0]), new FeedItem(X, test_X), new FeedItem(Y, test_Y));
            Console.WriteLine($"Testing cost={testing_cost}");

            var diff = Math.Abs((float)training_cost - (float)testing_cost);
            Console.WriteLine($"Absolute mean square loss difference: {diff}");
        }
    });
```

4.2 使用 TensorFlowSharp

TensorFlowSharp（https://github.com/migueldeicaza/TensorFlowSharp）提供了在 C#程序中可以使用的 API。这些 API 特别适合加载在 Python 中创建的模型并在.NET 应用程序中执行它们。

从 NuGet.org 获得的 NuGet 包附带了适用于 Windows（x64）、Mac（x64）和 Linux（x64）的本机 TensorFlow 运行时。

如果希望在其他平台上运行 TensorFlowSharp，可以通过为自己的平台下载相应的 TensorFlow 动态库并将其并排放置在 TensorFlowSharp.dll 库中来实现。

为了使用 TensorFlowSharp，需要在 Windows 上创建.NET 桌面应用程序或在 Linux 和 Mac 上使用 Mono。

请确保从 NuGet 下载 TensorFlowSharp 软件包，使用命令行（nuget install TensorFlowSharp）或从一个.NET IDE 下载软件包。在程序包管理器控制台输入：

```
PM> Install-Package TensorFlowSharp -Version 1.13.0
```

应用程序通常会创建一个图形（TFGraph）并在那里设置运算，然后从中创建一个会话（TFSession），然后使用会话运行器设置输入和输出并执行管道。像这样：

```
using(var graph = new TFGraph ())
{
    graph.Import (File.ReadAllBytes ("MySavedModel"));
    var session = new TFSession (graph);
    var runner = session.GetRunner ();
    runner.AddInput (graph ["input"] [0], tensor);
    runner.Fetch (graph ["output"] [0]);

    var output = runner.Run ();

    //从输出中获取结果
    TFTensor result = output [0];
}
```

在不单独设置图表的情况下，会话将为您创建一个图表。以下示例显示如何用 TensorFlow 来计算两个数字的相加：

```
using (var session = new TFSession())
{
    var graph = session.Graph;

    var a = graph.Const(2);
    var b = graph.Const(3);
    Console.WriteLine("a=2 b=3");

    //两个常量相加
    var addingResults = session.GetRunner().Run(graph.Add(a, b));
    var addingResultValue = addingResults.GetValue();
    Console.WriteLine("a+b={0}", addingResultValue);

    //两个常量相乘
    var multiplyResults = session.GetRunner().Run(graph.Mul(a, b));
    var multiplyResultValue = multiplyResults.GetValue();
    Console.WriteLine("a*b={0}", multiplyResultValue);
}
```

4.3 本章小结

本章介绍了通过 TensorFlow.NET 和 TensorFlowSharp 使用 TensorFlow 的方法。

第 5 章
Slurm 并行训练

高性能计算（HPC）通常指的是以一种方式聚合计算能力的实践。该方式提供比从典型的台式计算机或工作站获得的性能高得多的性能，以便解决科学、工程或商业中的大问题。随着数据集的增长，高性能计算对训练模型变得越来越重要。

本章首先介绍如何使用网格计算引擎 Slurm 构建 Linux 高性能计算集群，然后介绍如何实现 TensorFlow 在 Slurm 集群的运行。

5.1 网格计算引擎 Slurm 简介

Slurm（Simple Linux Utility for Resource Management）是一个开源、容错、高度可扩展的集群管理和作业调度系统，适用于大型和小型 Linux 集群。不需要为了使用 Slurm 而修改操作系统内核。作为集群工作负载管理器，Slurm 有三个关键功能：首先，它在一段时间内为用户分配对资源（计算节点）的独占和/或非独占访问，以便执行程序；其次，它提供了一个框架，用于在分配的节点集上启动、执行和监视（通常是并行作业）；最后，它通过管理待处理工作的队列来仲裁资源争用。

Slurm 已经使用 arm64（aarch64），ppc64 和 x86_64 架构在大多数流行的 Linux 发行版上进行了全面测试。目前支持的 Linux 发行版包括 Cray Linux Environment、Debian、RedHat Enterprise Linux、CentOS、Scientific Linux、SUSE Linux Enterprise Server、Ubuntu。

5.1.1 安装 Slurm

可以通过 Docker 的 Slurm 镜像来使用 Slurm 集群。

如果操作系统还没有安装应用容器引擎 Docker，则首先安装 Docker：

```
$ sudo apt install docker.io
```

https://hub.docker.com/r/giovtorres/docker-centos7-slurm/tags/提供了多个可用的标签。要使用最新的可用镜像，请运行：

```
#docker pull giovtorres/docker-centos7-slurm:latest
```

进入容器内的 bash shell。

```
#docker run -it -h ernie giovtorres/docker-centos7-slurm:latest
```

测试 slurmd 配置：

```
#slurmd -C
NodeName=ernie CPUs=16 Boards=1 SocketsPerBoard=2 CoresPerSocket=8 ThreadsPerCore=1
RealMemory=64393
```

supervisord 是一个进程管理工具。要查看所有进程的状态，请运行：

```
[root@ernie /]#supervisorctl status
munged                           RUNNING   pid 23, uptime 0:02:35
mysqld                           RUNNING   pid 24, uptime 0:02:35
slurmctld                        RUNNING   pid 25, uptime 0:02:35
slurmd                           RUNNING   pid 22, uptime 0:02:35
slurmdbd                         RUNNING   pid 26, uptime 0:02:35
```

查看配置文件 slurm.conf：

```
#cat /etc/slurm/slurm.conf
```

显示结果如下：

```
#集群名
ClusterName=linux
#主控制器的主机名
ControlMachine=ernie
#ControlAddr=
#BackupController=
#BackupAddr=
#slurm 进程用户
SlurmUser=slurm
#SlurmdUser=root
#slurmctld 控制器端口
SlurmctldPort=6817
```

```
#slurmd 节点守护进程端口
SlurmdPort=6818
#slurm 通信认证
AuthType=auth/munge
#JobCredentialPrivateKey=
#JobCredentialPublicCertificate=
#slurm 任务状态保存目录
StateSaveLocation=/var/lib/slurmd
#slurmd 守护进程日志保存
SlurmdSpoolDir=/var/spool/slurmd
SwitchType=switch/none
MpiDefault=none
#slurmctld 的 pid 存放
SlurmctldPidFile=/var/run/slurmd/slurmctld.pid
#slurmd 守护进程的 pid 文件存放
SlurmdPidFile=/var/run/slurmd/slurmd.pid
ProctrackType=proctrack/pgid
#PluginDir=
CacheGroups=0
#FirstJobId=
ReturnToService=0
#MaxJobCount=
#PlugStackConfig=
#PropagatePrioProcess=
#PropagateResourceLimits=
#PropagateResourceLimitsExcept=
SlurmctldTimeout=300
SlurmdTimeout=300
InactiveLimit=0
MinJobAge=300
KillWait=30
Waittime=0
#
#调度
SchedulerType=sched/backfill
#SchedulerAuth=
#SchedulerPort=
#SchedulerRootFilter=
SelectType=select/cons_res
SelectTypeParameters=CR_CPU_Memory
FastSchedule=1
#PriorityType=priority/multifactor
#PriorityDecayHalfLife=14-0
#PriorityUsageResetPeriod=14-0
#PriorityWeightFairshare=100000
```

```
#PriorityWeightAge=1000
#PriorityWeightPartition=10000
#PriorityWeightJobSize=1000
#PriorityMaxAge=1-0
#
#日志
SlurmctldDebug=3
#slurmctld 控制器守护进程的日志存放
SlurmctldLogFile=/var/log/slurm/slurmctld.log
SlurmdDebug=3
SlurmdLogFile=/var/log/slurm/slurmd.log
JobCompType=jobcomp/none
#JobCompLoc=
#
#审计
#JobAcctGatherType=jobacct_gather/linux
#JobAcctGatherFrequency=30
#
AccountingStorageType=accounting_storage/slurmdbd
#AccountingStorageHost=localhost
#AccountingStorageLoc=
#AccountingStoragePass=
#AccountingStorageUser=
#
#计算节点
GresTypes=gpu
NodeName=c[1-5] NodeHostName=localhost NodeAddr=127.0.0.1 RealMemory=1000
NodeName=c[6-10] NodeHostName=localhost NodeAddr=127.0.0.1 RealMemory=1000 Gres=gpu:titanxp:1
#
#分区
PartitionName=normal Default=yes Nodes=c[1-5] Priority=50 DefMemPerCPU=500 Shared=NO MaxNodes=5 MaxTime=5-00:00:00 DefaultTime=5-00:00:00 State=UP
PartitionName=debug Nodes=c[6-10] Priority=50 DefMemPerCPU=500 Shared=NO MaxNodes=5 MaxTime=5-00:00:00 DefaultTime=5-00:00:00 State=UP
```

在 slurm.conf 文件中，ControlMachine 主机名设置为 ernie。由于这是一体化安装，因此主机名必须与 ControlMachine 匹配。因此，必须在运行时将-h ernie 传递给 docker，以便主机名匹配。

如果对配置文件进行了少量更改，则可以使用 scontrol reconfig 命令让守护程序重新读取 slurm.conf。

可以运行命令 sinfo 报告由 Slurm 管理的分区和节点的状态：

```
[root@ernie /]#sinfo
PARTITION AVAIL  TIMELIMIT  NODES  STATE NODELIST
```

```
normal*      up 5-00:00:00      5   idle c[1-5]
debug        up 5-00:00:00      5   idle c[6-10]
```

scontrol 命令列出可以访问的分区:

```
[root@ernie /]#scontrol show partition
PartitionName=normal
    AllowGroups=ALL AllowAccounts=ALL AllowQos=ALL
    AllocNodes=ALL Default=YES QoS=N/A
    DefaultTime=5-00:00:00 DisableRootJobs=NO ExclusiveUser=NO GraceTime=0 Hidden=NO
    MaxNodes=1 MaxTime=5-00:00:00 MinNodes=1 LLN=NO MaxCPUsPerNode=UNLIMITED
    Nodes=c[1-5]
    PriorityJobFactor=50 PriorityTier=50 RootOnly=NO ReqResv=NO OverSubscribe=NO
PreemptMode=OFF
    State=UP TotalCPUs=5 TotalNodes=5 SelectTypeParameters=NONE
    DefMemPerCPU=500 MaxMemPerNode=UNLIMITED
```

如果需要从头开始安装 Slurm, 则首先要确保时钟, 用户和组 (UID 和 GID) 在群集中同步。Slurm 需要使用 MUNGE (https://dun.github.io/munge/) 来认证, 所以要先来安装 MUNGE。

```
#sudo apt-get install munge
```

确保群集中的所有节点都具有相同的 munge.key。确保在启动 Slurm 守护程序之前启动了 MUNGE 守护程序。

启动 MUNGE 守护进程:

```
#/etc/init.d/munge start
[ ok ] Starting munge (via systemctl): munge.service.
```

可以执行以下步骤来验证软件是否已正确安装和配置:

在 stdout 上生成凭证:

```
#munge -n
```

检查凭证是否可以在本地解码:

```
#munge -n | unmunge
```

检查凭证是否可以远程解码:

```
#munge -n | ssh <somehost> unmunge
```

如果遇到问题, 请检查 munged 守护程序是否正在运行:

```
#/etc/init.d/munge status
```

安装数据库 MariaDB:

```
#sudo apt-get install mariadb-server mariadb-client
```

以 MariaDB root 用户身份登录。

```
#sudo mysql -u root -p
```

输入密码后，将进入 MariaDB shell。

如果要启动 MariaDB，可以使用以下命令。

```
#sudo systemctl start mariadb
```

可以使用以下命令停止 MariaDB：

```
#sudo systemctl stop mariadb
```

创建目录：

```
#mkdir -p /storage
```

下载源代码：

```
#wget https://download.schedmd.com/slurm/slurm-18.08.7.tar.bz2
```

解压缩：

```
#tar xvjf slurm-18.08.7.tar.bz2
```

编译代码：

```
#cd slurm-18.08.7/
#./configure   --prefix=/tmp/slurm-build   --sysconfdir=/etc/slurm   --enable-pam --with-pam_dir=/lib/x86_64-linux-gnu/security/ --without-shared-libslurm
```

编译 Slurm。

```
#make -j 4
#make contrib
#make install
```

安装 fpm：

```
#apt-get install ruby ruby-dev rubygems build-essential
#gem install --no-ri --no-rdoc fpm
```

用 fpm 制作安装包：

```
#fpm -s dir -t deb -v 1.0 -n slurm-18.08.7 --prefix=/usr -C /tmp/slurm-build .
```

用 dpkg 命令安装 deb 包：

```
#dpkg -i slurm-18.08.7_1.0_amd64.deb
```

测试 slurmd 配置：

```
#slurmd -C
```

使用 https://slurm.schedmd.com/configurator.easy.html 上的在线工具，您可以生成 slurm.conf 配置文件。

在主节点启动控制守护进程：

```
#slurmctld -c
```

在计算节点上启动 slurm 守护进程：

```
#slurmd -c
```

5.1.2　Slurm 脚本编程

使用 srun 命令直接提交适用于快速简单的单个任务。例如，在 3 个节点上执行/bin/hostname，并输出任务编号。

```
#srun -N3 -l /bin/hostname
```

输出结果如下：

```
0: ernie
2: ernie
1: ernie
```

这里使用默认分区。默认情况下，每个节点一个任务。srun 命令有许多选项可用于控制分配的资源以及如何在这些资源之间分配任务。

使用 sbatch 命令将批处理脚本提交给 Slurm。显示 sbatch 帮助信息：

```
#sbatch --help
```

例如，批处理脚本 my.script 内容如下：

```
adev0: cat my.script
#!/bin/sh
#SBATCH --time=1
/bin/hostname
srun -l /bin/hostname
srun -l /bin/pwd
```

此脚本包含嵌入其自身作业的时间限制。可以通过使用前缀"#SBATCH"后跟脚本开头的选项（在脚本中执行任何命令之前）来根据需要提供其他选项。例如，如果需要重新启动失败的工作作业，可以使用--requeue 选项。

```
#SBATCH --requeue              ###失败时，重新排队进行另一次尝试
```

在节点 adev0 上提交作业：

```
adev0: sbatch -n4 -w "adev[9-10]" -o my.stdout my.script
```

```
sbatch: Submitted batch job 469
```

在此示例中，脚本名称为 my.script，声明使用节点 adev9 和 adev10（-w "adev[9-10]"，请注意使用节点范围表达式）。还明确声明后续作业步骤将分别产生 4 个任务，这将确保分配包含至少 4 个处理器（每个任务要启动一个处理器）。输出将出现在文件 my.stdout（-o my.stdout）中。

my.script 包含在分配中的第一个节点上（脚本运行的位置）执行的命令/bin/hostname，加上使用 srun 命令发起的两个作业步骤并按顺序执行。

使用 squeue 命令查看作业队列中作业的信息。例如：

```
#squeue
         JOBID PARTITION     NAME     USER ST       TIME  NODES NODELIST(REASON)
```

使用 scancel 命令取消已经提交的作业。例如：

```
#scancel 123456
```

Slurm 不会自动将可执行文件或数据文件迁移到分配给作业的节点。文件必须存在于本地磁盘或某些全局文件系统（例如 NFS 或 Lustre）中。Slurm 提供工具 sbcast，使用 Slurm 的分层通信将文件传输到分配节点上的本地存储。

Toil 是一个完全用 Python 编写的可扩展、高效、动态的跨平台管道管理系统。Toil 曾用于在 4 天内使用 32000 个可抢占计算核心的商业云集群处理 20000 多个 RNA-seq 样本。Toil 还可以运行 Common Workflow Language 中描述的工作流，并支持各种调度程序，例如 Mesos、GridEngine、LSF、Parasol 和 Slurm。

5.2 TensorFlow 集群

为了让 TensorFlow 运行在 Slurm 集群，可以使用 sbatch 命令将批处理脚本简单地传递给 Slurm：

```
#sbatch --partition=part start.sh
```

可以使用 sinfo 列出可用分区。

start.sh 中可能的配置如下：

```
#!/bin/sh
#SBATCH -N 1              #请求的节点
#SBATCH -n 1              #请求的任务
#SBATCH -c 10             #请求的 CPU 核
#SBATCH --mem=32000       #以 MB 为单位的内存
#SBATCH -o outfile        #将 stdout 发送到 outfile 文件
```

```
#SBATCH -e errfile  #将stderr发送到errfile文件
python run.py
```

这里的 run.py 包含想用 slurm 执行的脚本，即 TensorFlow 代码。

5.3 本章小结

本章介绍了如何使用网格计算引擎 Slurm 构建 Linux 高性能计算集群和如何实现 TensorFlow 在 Slurm 集群的运行。

参考文献

[1] 罗刚，张子宪，崔智杰. Java中文文本信息处理——从海量到精准[M].北京：清华大学出版社，2017.

[2] 罗刚. 使用C#开发搜索引擎[M]. 2版.北京：清华大学出版社，2018.

[3] 罗刚.网络爬虫全解析——技术、原理与实践[M].北京：电子工业出版社，2017.

[4] 柳若边. 深度学习：语音识别技术实践[M].北京：清华大学出版社，2019.